BLAC MARKERS

EDINBURGH'S DARK HISTORY
TOLD THROUGH ITS CEMETERIES

JAN-ANDREW HENDERSON
PHOTOGRAPHS BY JAMIE AND CHARLOTTE CORSTORPHINE

AMBERLEY

For Char
Who worked out where all the markers were.

Scotland has a national treasure which has not yet received just recognition. That treasure is contained in the kirkyards...

Stones. Betty Willsher and Doreen Hunter (1978)

To write the story of the relics of humanity that have a last resting place here, is probably to indite a history of the city itself, if not actually the whole country.

History of the Old The Greyfriars Church. W. Moir Bryce (1912)

First published 2015

Amberley Publishing
The Hill, Stroud
Gloucestershire, GL5 4EP

www.amberley–books.com

Copyright © Jan–Andrew Henderson, 2015

The right of Jan–Andrew Henderson to be identified as the Authors of this work has been asserted in accordance with the Copyrights, Designs and Patents Act 1988.

British Library Cataloguing in Publication Data.
A catalogue record for this book is available from the British Library.

ISBN 978 1 4456 4798 2 (print)
ISBN 978 1 4456 4799 9 (ebook)

Typesetting and Origination by Amberley Publishing.
Printed in Great Britain.

Contents

Prologue

Let's go looking for buried treasure.

The treasure is Edinburgh's dark and fascinating history, which is not as well known as you might expect. The majority of the city's own residents only know a few of the most famous tales and visitors to the capital get most of their information from the top of a tour bus. Admittedly, given Edinburgh's notorious traffic congestion, this can be quite in–depth, until the guide runs out of script and begins to tell you what he had for lunch.

There's so much more to our past. Edinburgh is now a progressive modern metropolis, but remained nasty and medieval long after most other European cities had cleaned up their act. Once a cauldron of religious fanaticism and persecution, it was a lynchpin in the age old conflict between Scotland and England. It sent innovators to the four corners of the earth and eventually gained fame as 'The Athens of the North', leading the world in philosophy, sciences, medicine, architecture, art and literature. At the same time, Edinburgh was a den of iniquity, astonishing poverty and horrendous crime. A number of historians think Robert Louis Stevenson's masterpiece *The Strange Case of Dr Jekyll and Mr Hyde* is partly an allegory about the place where he grew up. The Jekyll and Hyde city. Who wouldn't want to read the story of somewhere like that?

In that spirit, this book is not just a long list of dates and facts about the past. I want to tell a (hopefully) vibrant, enthralling tale about a vibrant, enthralling town. I'm an ex–tour guide myself, so I know that how a story is presented makes all the difference between getting a tip or something thrown at me. I won't even mention what I had for lunch. Though it was a cheese sandwich, in case you're interested.

So where does the buried part come in?

It's because Edinburgh's graveyards provide all the clues. They contain monuments galore and it's where many of the movers and shakers are buried, both heroes and villains. Some are famous. Some totally forgotten.

I intend to bring them back to life. Not literally, because that would be creepy and get me arrested. Instead, I want to weave them into a tapestry that gives a dramatic picture of Edinburgh's past. Besides, each cemetery has captivating stories of its own. Greyfriars Graveyard, for instance, was the launch pad for the English Civil War and the subsequent restoration of the monarchy; it contains an early concentration camp and was even used as a racetrack. In my opinion, it inspired Darwin's Theory of Evolution, the novel *Frankenstein*, the musical *Cats*, the *Harry Potter* books and is the lair of the world's best documented supernatural case.

People visit graves, after all, because they feel it brings them closer to the departed. That's what I want to achieve on paper, by providing a bit of context to the times in which those extinguished lives were lived. I'll only be scratching the surface, if you'll excuse the pun, but I've tried to cover as much interesting historical ground as possible, while charting Edinburgh's evolution from ancient times to the present.

So, I won't be taking each graveyard in turn. Instead I'll dip in and out of several representative sites to build up a picture of what Edinburgh used to be like. I have included, however, a list of notable people and monuments in each burial ground at the end, if you care to take a look for yourself.

I sincerely hope you do.

Jan–Andrew Henderson

Ancient Times

Long ago is tricky when you're telling the story of a city using its graveyards, for many of Edinburgh's ancient burial grounds are now gone. Though we know their approximate location, there isn't much left to see, for they're usually underneath a modern structure. St Giles graveyard on the Royal Mile is now a car park and its most famous resident, John Knox, has a little stone plaque marking his last resting place in bay 23. After almost four and a half centuries, it's unlikely to be in the right spot, which is probably for the best, since there's usually a car parked on top.

Others, like St Cuthbert's, bear little resemblance to their original incarnations or, in the case of Holyrood Abbey, lie in ruins. Yet all have a tale to tell.

When we come to proper memorials there are all sorts – headstones, tombs, plaques, flat gravestones, obelisks, sarcophagi and monuments, to name a few. To keep things simple, I will refer to them all as 'markers', unless otherwise stated. In the same way, I have used the word 'graveyard' to refer to any burial ground. Technically, a graveyard has a church attached while a cemetery does not. A cemetery also allows cremated bodies to be interred there.

Written records of Edinburgh's ancient history are as thin on the ground as physical ones, but I'll give a rough overview. Traces of Mesolithic, Iron Age and Bronze Age settlements have been discovered in the area and first century Roman invaders found a Celtic tribe, the Votadini, living there. By the seventh century they had been replaced by the Goddodin, which sounds like a name J. R. R. Tolkein made up. They called their home *Din Eidyn* or *Etyn* (the spelling varies, depending upon sources), which eventually morphed into the Gaelic *Dun Eideann* and finally became Edinburgh. 'Edin' is an old word for slope and the Old Town stands on a glaciated ridge.

It's great stories and the people they are about that bring history to life, so that's what I'll concentrate on. By a happy coincidence, Edinburgh doesn't start to come into its own as an amazing city that shaped world events until the mid–sixteenth century, at which time graveyards in the form we see today began appearing. Before that, the rich were interred inside parish churches and more common people buried in the surrounding land, sometimes referred to as 'God's Acre'. Virtually none of these markers have survived but I can't just sweep several centuries under the carpet. So here are the highlights.

In AD 683, Edinburgh was overrun by the Angles of Northumbria, at that time a kingdom to be reckoned with. It remained under the Angles control until the tenth century, when it was abandoned to the increasingly influential Scots, who had crossed the sea from Ireland. Though he was based in Dunfermline, King Malcolm Canmore (1058–93) and his wife Margaret began to spend more time in Edinburgh and subsequent monarchs followed suit. If you've ever been to Dunfermline, you'll understand why.

In the twelfth century Margaret's son, David I (1084–53), awarded the town Royal Burgh status and built a chapel commemorating his mum in what is now Edinburgh Castle. St Margaret's Chapel stands to this day, the oldest surviving building in the city.

In 1128, David also founded Holyrood Abbey, though it was to be demolished, rebuilt, damaged and remodelled several times. Adjacent buildings were often used as royal lodgings and the Scottish parliament sat there several times between the thirteenth and fifteenth century, including one presided over by Robert the Bruce. In 1768, the decaying roof finally collapsed, leaving an impressive skeleton that still stands today.

In 1829, the composer Felix Mendelssohn cited the Abbey as an inspiration for one of his most famous works:

> The chapel is now roofless, and is overgrown with grass and ivy, and the ruined alter where Mary Queen of Scots was married. Everything is in ruins and mouldering, and the bright light of Heaven shines in. I believe I have found the beginning of my Scottish Symphony here today.

Fortunately, it is no longer mouldering but in pristine condition, probably because you have to pay to get in.

Though some of the characters in our story are interred there, age, sackings and renovations mean many of their markers are gone. The Abbey was also the site of several royal interments, including David II, James II and James IV, whose tombs were later moved to the 'Royal Vault' in the south aisle.

Medieval Edinburgh continued to rise in prominence, enduring several upheavals on the way. Everyone knows about William Wallace, Robert the Bruce and the thirteenth century Scottish Wars of Independence but they came and went without really disrupting the town's growth. Despite Edinburgh changing hands between the Scots and English a couple of times, towns like Glasgow and Perth played far greater roles in the conflict. Bruce's son, David II (1324–71) may have seen the city's potential however, for he spent years fortifying the castle's defences, including the sixty–feet–high King David's Tower.

Edinburgh emerged from the so–called Dark Ages firmly established and could look forward to many uninterrupted centuries of war, plague, religious upheaval, rebellion, poverty and crime. The town's many burial grounds were set around priories, friaries and churches, like St Giles, Blackfriars, St Catherine's, St Roque and the Kirk O' Fields – the latter now underneath Chambers Street and the National Museum. Today, they are only evocative names, victims of the march of time.

Occasional finds do pop up now and then. The remains of a medieval knight were excavated among the ruins of a thirteenth–century Dominican friary under yet

Holyrood Abbey
and burial ground.

The Royal Vault,
Holyrood.

another car park, appropriately belonging to the former archaeology department of Edinburgh University. What is it with the dead and parking lots? Even Richard III ended up in one.

As for the existing graveyards we will look at, two are far older than the others, so let's visit them first.

St Cuthbert's and Colinton

St Cuthbert's is now at the west end of Princes Street, with the magnificent backdrop of Edinburgh Castle rising up behind. Of course, it was all fields back then, as old folk like to say. That's because there has been a chapel in some form on the site as far back as the eighth century, a thousand years before Princes Street was even conceived. It seems likely that there was a graveyard even then, making St Cuthbert's burial ground many centuries older than the others we will explore. Though the original graves are gone and there are no monuments predating the Reformation, there is still plenty to capture the imagination.

Like St Cuthbert's church itself, the burial ground has undergone many upheavals. It was extended in the eighteenth century and, in 1841, a railway tunnel was built underneath, resulting in the destruction of many markers. All that is left of the original graveyard is a small mound known as 'Bairn's Knowe', because children were often interred there. Infant mortality rates being what they were in ancient Edinburgh, I'm surprised it's not the size of Everest.

Today, St Cuthbert's is ringed by Lothian Road and Princes Street, both looking down into a morass of higgledy–piggledy markers, bent boughs and terraces lined by mossy walls. Each are studded with weather beaten monuments and mausoleums, dark with age. As soon as you descend the steps, traffic disappears and you find yourself in an altogether more tranquil, faintly melancholy valley of the dead. It also has some fine Memento Mori, or reminders of our mortality.

You might think standing in a graveyard is a perfectly adequate reminder of your inescapable fate but latter day Scots weren't a particularly subtle bunch.

Colinton church was established in the late eleventh century (though it was replaced around 1650) so there has been some form of graveyard there for almost a thousand years. It's a well hidden site and certainly one of the most idyllic and picturesque graveyards in the city.

Next to the former village of Colinton, sheer wooded cliffs rise up on one side, while the Water of Leith meanders past below. The burial site isn't large but still contains a few notable residents, like James Gillespie (1726–97), a tobacco merchant who founded James Gillespie's School. Great combination tobacco and schools, but I have to include him because my son goes there. Gillespie is also credited with originating the phrase 'Waste not, want not', which is as typically Edinburgh as you can get.

St Cuthbert's Graveyard with Edinburgh Castle in the background.

Memento Mori, St Cuthbert's.

Colinton Graveyard.

The notable palaeontologist Ramsey Heatley Traquair (1840–1912) is interred there, as is his wife, Phoebe Traquair (1852–1936), leading light of the Scottish Arts and Crafts movement. Check out her murals in St Mary's Cathedral or the former Catholic Apostolic Church on East London Street (now the Mansfield Traquair Centre), which was called 'Edinburgh's Sistine Chapel'. She even designed her own headstone.

Like all of Edinburgh's burial grounds, Colinton has some great tales attached. One of its ministers was Dr Lewis Balfour, maternal grandfather of Robert Louis Stevenson, and the famous writer spent a great deal of time there as a boy. He was a literary magpie when it came to Edinburgh locations (A mound that can be seen from his boyhood house became *Treasure Island,* while the lamppost outside inspired his well–known poem *The Lamplighter*), and the graveyard was no exception. Next to the church is a 500–year–old yew tree, which still has the metal fittings for the swing that young Stevenson played on.

This was almost certainly the basis for his poem 'The Swing' from a 'Child's Garden of Verses'.

> How do you like to go up in a swing
> Up in the air so blue?
> Oh, I do think it the pleasantest thing
> Ever a child can do!

After that peaceful interlude, we shall roll back the centuries and return to the dark history of Edinburgh.

Ramsey and Phoebe Traquair, Colinton.

James Gillespie's
mausoleum, Colinton.

Stevenson's swing,
Colinton.

The Early Stuarts

In 1371, David II died without an heir and the throne passed to his nephew Robert II, the first monarch of the Stuart Dynasty. We need to know about the Stuarts because their royal line would play a huge role in Edinburgh's story, for better or worse.

The Stuarts were the most famous of Scotland's monarchs and debate rages about whether they were good or bad, clever or stupid. Of course, the further back we go, the harder it is to distinguish truth from legend. So the following is a mixture of both.

Scotland was a truly violent place when the Stuarts came to power. Beleaguered by his own nobles, the sickly and ineffectual Robert III (1337–1406) sent his young son, James, to France for safety. James was captured by the English on the way and Robert was so upset he promptly died. According to the fifteenth–century chronicler Walter Bower, his final request is the least pompous of anybody who has ever strutted the stage of history. He asked to be interred on a rubbish heap with the epitaph 'Here lies the worst of the Kings and the most wretched of men'.

Even his last wish wasn't granted, for he's buried in Paisley.

Actually, James had a fine time down south, even accompanying the English army against a combined Scots and French force. Released after eighteen years of captivity, he proved far less popular with the nobles of his own country and, in 1437, a group of them murdered him as part of a failed coup. When they broke into James' Perth residence, the king fled into an underground drain. He would have escaped but had ordered the other end sealed up because he kept losing tennis balls down there.

Good news for Edinburgh, however. The Stuarts finally made it their official capital, perhaps because they thought it might be safer than Perth. This shows how wrong you can be.

James II (1430–60) was a lively and active monarch with an unfortunate enthusiasm for modern artillery. At a siege in 1460, he was killed by an exploding cannon and buried in Holyrood. At least he went out with a bang.

James III (1451–88), was such an unpopular king that another rebellion began against him, again led by those pesky nobles, who put the king's own son at the head of their army. According to legend, James fell off his horse, probably in surprise at being faced by his own offspring. He was taken to a mill where, just in case his wounds were serious, asked for a priest. The miller's wife brought back an assassin disguised as a clergyman, who murdered him. That's country folk for you.

His son, James IV (1473–1513), grew up to be the greatest king Scotland ever had. He was charming and intelligent and under his rule Scotland looked set to flourish at last. So, typically, he threw it all away. To help the French in yet another pointless war he invaded England in 1513 and was killed in Scotland's greatest military disaster, the Battle of Flodden.

Fearing an immediate reprisal, Edinburgh Council ordered a defensive barricade built round the city. The 'Flodden Wall', however, turned out to be an utter liability. The anticipated English counter attack never came and, in future years, the wall spectacularly failed to stop any invasion. What it did was keep the citizens penned up inside for more than two centuries. In that time the population tripled until there were 30,000 people crammed into roughly 140 acres. This gave rise to the 'lands', tenements of up to eleven stories. When the builders couldn't get any higher, they dug down into the Old Town ridge instead, creating a subterranean warren known as the 'Underground City', the history of which I covered in my book *The Town Below the Ground*.

Far from optimum living conditions, you'll agree. Edinburgh famously stank and plague frequently broke out, many of the victims ending up buried in mass graves under Bruntsfield links. I suppose that qualifies the area as a graveyard of sorts, so I have included a picture. It also vies with Leith Links in Edinburgh as the place where golf was invented.

Little remains of the Flodden Wall today, but there is still an imposing part left in what is now Greyfriars Graveyard.

James V (1512–42) didn't learn from his father's mistakes. He also sent a huge army against a tiny English force and they were sent packing, which seems to have put him into a fatal decline. After hearing that his wife had given birth to a daughter rather than a male heir, he uttered the cryptic message, 'It began with a lass, it'll end with a lass.' Before anyone could work out what it meant, he turned his face to the wall and died. He is also buried in Holyrood.

Perhaps he had a premonition of what was to come, for his daughter was Mary Queen of Scots (1542–87).

Here is where our story of Edinburgh really takes off, as the city began carving out a unique identity, becoming a focal point of Scottish history. By happy coincidence, the next of our existing graveyards was established during Mary's reign and the monuments and markers you can actually see chart Edinburgh's progress. Though there are undeniable highlights, it's a bitter, uneasy tale.

The next home of our black markers had arrived.

Bruntsfield Links.

Flodden Wall, Greyfriars.

Greyfriars Graveyard

Greyfriars is built on the site of an old Franciscan Friary and is my personal favourite. I even lived in a flat overlooking the place.

Greyfriars opened in 1562, after St Giles' graveyard was declared full and the Grey Friars had been chased off to Holland by Protestant Reformers, who we will be hearing more of soon. Surrounded by high walls and tenements, it's almost invisible from the outside, despite being in the heart of Edinburgh.

It's a pleasant place in the daylight. Tourists wander between the trees and office workers eat packed lunches on flat gravestones. At night, the area takes on a different hue. Greyfriars has a fierce reputation for being haunted and is frequented by ghost tours. Some people object to this, but it has certainly got rid of the drug addicts and drunks that used to frequent the place.

This will be a recurring theme in Edinburgh's story. Social problem? Easier to move the people somewhere else than try to fix it.

Due to the large number of high quality gravestones sculpted by renowned masons and architects, Greyfriars is now considered a major resource of Scottish seventeenth and eighteenth–century artwork. Restrictions placed by the Deans of Guilds, from

Greyfriars
Graveyard.

the 1590s until the 1900s, determined the placement and style of the monuments and ensured only the richest and most prominent families could afford them. According to Grant's *Old and New Edinburgh*:

> Here lie thirty–seven chief magistrates of the City; twenty–three principals and professors of Edinburgh University; many of them of more than European Celebrity; thirty–three of the most distinguished lawyers of their day – one a vice–chancellor of England and Master of the Rolls, and another who was accountant general of the court of chancery; six lords president of the supreme court of Scotland; twenty–two senators of the college of justice, and a host of men distinguished for the splendour of their genius, piety and worth. Here too lie, in unrecorded thousands, citizens of a more humble position.

This is a bit of a brush off, if you were one of those 'humble' unrecorded citizens but that's another feature of Edinburgh's story we'll see repeated.

Greyfriars also has some of Scotland's oldest and best preserved 'Memento Mori'. Death is represented by The Grim Reaper, time, spears, bows and arrows, coffins and bells. Hourglasses represent the passing of time (a vertical hourglass meant someone who had lived a full existence and a horizontal one was a life cut short, as were skulls facing each other). Symbols of the dead person's trade were a common feature and Greyfriars has more carved skulls and skeletons – also reminders of death and mortality – than you can shake a scythe at.

Most importantly, Greyfriars is crammed to the gills with fantastic stories. Here are a couple of the most quirky.

According to legend, T. S. Elliot once rented a flat overlooking the graveyard. Spotting numerous moggies outside his window, he began to name them. This led to his famous *Old Possum's Book of Practical Cats*, which eventually became the musical *Cats*. Men in furry leotards and multi coloured fright wigs? Elliot has a lot to answer for.

Memento Mori,
Greyfriars.

For dog lovers, we have the nineteenth–century tale of Greyfriars Bobby, the faithful Skye terrier who sat on his master's grave for fourteen years. The dog's effigy outside is the most photographed statue in Edinburgh, despite looking nothing like contemporary pictures of him. His marker is also the first you see entering the graveyard and, without trace of irony, the city council have erected a sign opposite saying 'No Dogs Allowed'. I don't want to shatter illusions, but a historian friend named Jim Gilhooley ploughed painstakingly through the newspaper records of the time and discovered Bobby's owner, John Grey, was actually buried two miles away in East Preston Street. If he's right, the poor dog wasn't only sitting on the wrong grave but in the wrong cemetery entirely.

Why did he stay? Well, the pie shop next door fed him and what else was he going to with his time? He was a dog.

Left: Bobby's Statue, Greyfriars.

Right: John Grey's grave, Greyfriars.

Bobby's Grave, with Memento Mori in the background, Greyfriars.

The Reformation

In the sixteenth century, when Greyfriars was founded, Edinburgh was a truly miserable place to live. Overcrowded, disease–ridden, filthy and wreathed in smoke, the city's casual violence was as common as it was shocking. One notorious brawl between the Douglas and Hamilton families involved 500 men battling through the wynds and closes of the town. Around eighty Hamiltons were killed and the fight became known as 'The Cleansing of the Causeway', due to blood running down the High Street.

When Edinburgh did produce someone spectacular, it was more by luck than design. Take John Napier, who lends his name to Napier University and was lucky enough to be buried inside St Cuthbert's Church, before the Reformers put a stop to the practise. Napier, loathed by all schoolchildren for discovering Logarithms, was a physicist, astronomer and mathematician of rare genius. Yet a letter from an uncle to his father gives an insight into Edinburgh's lack of culture at the time.

John Napier's tomb, St Cuthbert's.

I pray you, sir, to send John to the schools either of France or Flanders, for he can learn no good at home.

Hardly a surprising attitude. In 1595 disaffected pupils took over the High School and shot dead the Baillie who came to quell their revolt. Being the spoilt offspring of rich parents, they didn't even get detention. Apparently, some things never change.

In fact, Scotland's education system was about to improve immensely, but a bloody price would be paid for that advancement.

The Reformation was about to begin.

At the beginning of the sixteenth century, Scotland was a mainly Catholic country, but a new force was sweeping Europe. Inspired by the German, Martin Luther, Scottish Protestantism was on the rise. Like many other nations, the Scots had become increasingly frustrated by what they saw as greed, corruption and hypocrisy among the Catholic clergy. It didn't help that Henry VIII of England had been excommunicated and decided, since England was now Protestant, Scotland should be too. Especially if it would sever their strong ties with his old enemy, Catholic France.

To achieve this, he proposed that the infant Mary Queen of Scots should marry his son Edward. When the Scots objected, Henry began the 'Rough Wooing'. Hampered not a jot by the Flodden Wall, he invading the country in 1544, burned a large part of Edinburgh to the ground and inflicted severe damage to Holyrood Abbey. Romance never was Henry's forte.

As usual, he didn't get the result he wanted. Mary was moved to the safety of France, grew up Catholic and married the future king, Francis II. So much for severing ties with the Auld Alliance.

While she was gone the Protestant movement began slowly but surely gaining support in Scotland, despite a clampdown by the Catholic Church, who even executed some of the so called 'heretics'. Regent Arran, who ruled in the young queen's place, however, was sympathetic to the new religion – allowing a foothold to become a movement.

In 1560, Mary was dealt a triple whammy. Her husband died and back in Scotland, so did her mother, who had been the Reformers' staunchest opponent. Since she was officially the monarch of Scotland, the eighteen year old Mary returned to rule a country she was completely unfamiliar with and didn't much like. The feeling was mutual, for Scotland had finally made a break with the Catholic Church, establishing its own Presbyterian Kirk and governing body, the General Assembly.

Worse than that, a nemesis was waiting for her.

John Knox.

Go round the back of St Giles' Cathedral and, as I've said, you'll find a little grey stone on parking lot 23, marking the last resting place of John Knox. It's hard to read because it's the same colour as the asphalt and it's highly doubtful Knox is still there. In 1879, the city authorities dug up several tons of remains, mixed them with limestone and relocated the lot to Greyfriars. So good luck finding him.

Knox is quite the character. Born near Edinburgh in 1505, he was actually ordained as a Catholic priest, before converting to Protestantism. In 1545, he became a bodyguard for the famous Reformer, George Wishart, eventually tried and burned as a heretic outside St Andrew's Castle and buried in Holyrood. So much for having Knox as a minder.

In protest, a Protestant faction captured the fortress, hoping Henry VIII would come to their aid. A French force turned up first and Knox spent the next nineteen months as a galley slave, which may well explain why he became such a fanatic. If you don't fervently believe God has a purpose for you, you've just wasted two years rowing around chained to a bunch of sweaty men for no good reason.

By 1559, the Protestants were strong enough for the exiled Knox to return and, a year later, he became minister of the most important Protestant Kirk in the country – St Giles' on Edinburgh's Royal Mile. He was just in time to witness Mary Queen of Scots take over the throne and, as you can imagine, was none too enthusiastic about it.

Personally, I'm on Mary's side. Though a staunch Catholic, she didn't attempt to try and convert her people back to their old religion. Yet, when she tried to hold her own private mass, there was a riot and John Knox almost had apoplexy. Knox, on the other hand, was an intolerant bigot who banned Christmas as being too Pagan and wrote *The First Blast of the Trumpet Against the Monstrous Regime of Women,* a title that sorely needs an editor's touch. In it he argued that female rulers were contrary to the Bible, yet his low opinion of women didn't stop him marrying a seventeen year old when he was fifty. This was a more common occurrence than today, but still a bit icky.

For all his harrumphing and chest beating, Knox didn't bring about Mary's downfall. She managed that all by herself.

John Knox plaque,
St Giles'.

Mary Queen of Scots

Mary's decline began with a bad marriage and her attachment to an Italian called David Rizzio, who is buried in Canongate Kirkyard.

The queen's first fatal mistake was marrying her cousin, Lord Darnley, a union the Scots heartily disapproved of. Darnley and Mary were both grandchildren of the previous English monarch, Mary Tudor, and the union strengthened their claim to be Elizabeth I of England's successor. But Darnley was also a hated Catholic, whose father played a major role in the Rough Wooing. Whether she was in love or overly ambitious, you have to remember Mary was still in her twenties, so I think we should cut her some slack.

When her husband turned out to be a debauched playboy, the pregnant queen sought solace with her private secretary, a musician and singer named David Rizzio. Though contemporary accounts describe Rizzio as short, ugly and hunchbacked, Darnley grew jealous. He even suspected Rizzio was the real father of their child, the future James I of Britain.

A heart to heart talk might have been more prudent. Instead, Darnley had Rizzio stabbed to death in front of the queen, not just to get rid of a rival, but possibly to cow his wife into submission. If so, he sorely underestimated her.

In 1567 he was found strangled in Kirk O' Field near Greyfriars and is buried in Holyrood. Nobody knows for sure who did it, but the Scots strongly suspected the Earl of Bothwell, who had his eye on the pretty young queen. Bothwell then tricked

David Rizzio,
Canongate.

Mary, citing threat of rebellion, into moving to his home at Dunbar castle and may even have raped her. Not a great time to be a woman, the sixteenth century.

Whether she feared him or simply wanted the protection of a powerful Protestant, Mary married Bothwell three months later. This only confirmed suspicions that she had been party to Darnley's murder, so the Scots rose up against the pair. To make matters worse, one of the rebellion's leaders was Mary's own half brother, James Douglas, fourth Earl of Morton.

Morton seems to have been a real piece of work, even by sixteenth–century standards. Ambitious and ruthless, he was one of the men who stabbed Rizzio to death and probably had a hand in Darnley's murder. His insurgence scared Bothwell so much, the queen's 'protector' took off for Scandinavia, never to return. Mary was imprisoned, forced to abdicate and her young son (only one year old), James, was placed on the throne.

Scotland was subsequently ruled by a succession of regents, who did their best to turn James against his mum and make sure he grew up a strict Presbyterian. This turned out to be only partially successful, for he eventually moved Mary's body to a marble tomb in Westminster Abbey and espoused Episcopalinism rather than Presbyterianism.

Morton was the last of Scotland's regents, a position he was undoubtedly angling for all along. When Mary escaped from captivity and attempted to regain her throne, he led the army that defeated her, which is sibling rivalry on an epic scale. The queen was forced to flee to England where her cousin, Elizabeth I, incarcerated her again and eventually executed Mary in 1587. Which just goes to show, you should never trust relatives if you're famous and prettier than they are.

Morton's fate was an equally unhappy one, not that he didn't deserve it. When the nobles tired of his autocratic rule, he eventually faced his own rebellion and was executed for Darnley's murder in 1581.

It's a sordid story that nobody comes out of well. However, it's worth taking a closer look at Rizzio's and Morton's markers again.

James Douglas,
Greyfriars.

Rizzio was originally buried in Holyrood Abbey then, 120 years later, reinterred in Canongate Kirkyard. It seems rather odd that a Catholic who was hated when he was alive and had no surviving family would end up with a prominent marker in a Protestant Kirkyard, so I strongly suspect he's still in Holyrood. Still, this is Scotland. Pay me enough money and I'll arrange to have you buried at the bottom of my garden.

Plus Rizzio appears to be a persistent fellow. There is a rusty stain on the floor of Holyrood palace which is reputed to be his blood (he was stabbed fifty–six times), even though the floor has been replaced twice.

You'd think our present queen could afford better cleaners.

Early Markers

Morton's grave also has a peculiar story. Take a look at the marker again. It's pretty paltry for a man who once ruled a nation and, if genuine, should really read J. D. for James Douglas instead of J. E. M. (James, Earl of Morton). Besides, executed criminals, no matter their former status, were buried in unmarked graves in another part of Greyfriars. Though Morton is in the cemetery somewhere, it's probably not where his marker stands. Maybe it's better that way, for the sake of his dignity. The original stone is supposed to have sunk into the soggy ground and the shape of the present one means it gets used as a seat by tourists.

Morton is not the only denizen to suffer such a misfortune. The marker of the next notable to be buried in Greyfriars, historian and scholar George Buchanan, also sank without trace. He now has two replacements, which is a bit excessive, but the earliest was erected in the mid–nineteenth century. By that time, I suspect the precise whereabouts of poor George, who died in 1582, was simply a guess.

This is a pattern that will repeat itself over and over. Annoyingly, many of the older interments chronicled in this book may not be where they are signposted, as the present markers were erected long after they died. Short of a zombie apocalypse, however, we'll just have to work with what we have.

Back to the Reformers. Once in power, they passed an edict forbidding burials inside the churches, which traditionally held the aristocracy, members of guilds and any other professionals who could afford to pay for the privilege. This meant the land outside filled all the quicker, so the sites were expanded and gardens of Catholic monasteries and friaries (once the occupants had been frightened away) were converted into graveyards like Greyfriars. The Reformers also sacked Holyrood Abbey and destroyed any effigies and stained glass windows in Edinburgh's existing churches, as they were symbols of the hated Catholics.

Until the late sixteenth century, as I have already bemoaned, Scottish graveyards had virtually no markers. This was partly because they were used for many other purposes, such as archery practise, weapons drills, fairs, markets and general social gatherings. Funerals of the time must have been a rather anarchic affair. 'Watch where you're shooting those arrows, mate. I'm trying to bury someone here!'

When monuments did start to go up, it evidently annoyed the locals. If you look hard enough, you'll find that some markers have little holes where they've been used as target practice.

The rich and famous got round this problem by choosing the walls of burial grounds as a buttress for their imposing memorials, where they could be gaped at without getting in the way of resentful vandals. When markers eventually spread across the

George Buchanan
memorial,
Greyfriars.

rest of the graveyards, they retained a rigid social etiquette, outlined by the historian J. Weever in 1631.

> Sepulchres should be made according to the quality and degree of the persons deceased … persons of … plebeian sort shall be buried without any tomb or gravestone or epitaph; persons of the meanest sort of gentry a flat gravestone; Gentlemen of more emenencie … effigies and representations raised aloft and their personages delineated … epitaphs were only for such as were of virtue, wisdom and valour.

So there you go. If you were a plebeian, not only was your life miserable but you weren't even worth being remembered.

There was one positive outcome of the Reformation, a zeal to provide universal education for all classes in Scotland. Schools began to spring up in Edinburgh, laying the foundation for an astonishing transformation in later centuries. The Reformers also advised that these establishments be located near to graveyards, so the children would have a place to play on their breaks. Practical bunch, the Presbyterians.

Yet kids will be kids, even in those days. If you look in some of the more fancy mausoleums in Greyfriars you can see where pupils and workmen from long gone eras have carved their names into the stone tombs.

Wall tombs, Greyfriars.

Ancient graffiti, Greyfriars.

The Union of the Crowns

At this point in our story, Scotland was officially one nation. However, that's not strictly accurate, for the north was another matter entirely.

It's important to state that, when I'm talking about Scotland, I actually mean southern Scotland – where Edinburgh is located. I'll refer to northern Scotland as the highlands, with good reason. While the south had become Protestant, the north was still fiercely Catholic. The highlanders retained the old clan system, based on ancient family territories and customs and frequently raided their lowland countrymen. They were regarded with animosity and fear by their neighbours, who considered them savage, unpredictable and disturbingly ginger haired. Eventually the country would be truly united but that's still a long way off.

When it came to headstones, there was also a big difference. Post–Reformation markers in the highlands tended to have just dates and initials. In the lowlands they were more elaborate and contained actual information, luckily for us.

Approaching the seventeenth century, the future for Edinburgh seemed rosy. The Catholics had lost and the unstable rule of the regents was over. Mary's son, James VI, managed to reach maturity without getting killed and seemed an able enough ruler. Even better, when Elizabeth I of England died without an heir in 1603, he became king of both countries. The Scottish nobles must have rubbed their hands in glee. Their monarch now ruled England too. Result!

But James wasn't the sentimental sort. Though born in Edinburgh, he promptly took off for London, a move he described as 'swapping a stony couch for a deep feather bed'. Since Scotland was still in the grip of staunch Presbyterianism, which frowned on things like dancing, singing and smiling, he took his court poets, artists and musicians with him. Edinburgh suddenly lost her status as centre of the arts and when the city eventually became the hub of culture again, that culture would have a very different emphasis.

James promised to return frequently but, despite ruling for twenty–two years, only came back once – to hunt deer and possibly collect a few personal knick–knacks.

The new king had some odd quirks, including a conviction that witches were out to get him. So he reciprocated by instigating a whole raft of witch trials. There is a monument to all those who died on Edinburgh's Castle Hill, a popular place for burning them. It isn't where any are buried, but I feel it deserves inclusion, as most visitors don't even notice it's there.

Despite his penchant for murdering harmless old biddies, James was far from a terrible king. Yet there were deep differences between Scotland and England that the union of the crowns only highlighted.

One division was the fact that Protestantism had evolved into separate entities in each country – the Church of Scotland (Presbyterian) and the Church of England (Anglican). Though the Anglicans still had traces of Catholic ceremony, including bishops, the Scottish Presbyterian church was far more severe and considered Anglicans too close to the despised Catholics for their liking.

The second division was economic. England was prosperous. Scotland was poor, cold, wet and plagued by plague and infested with midges. Though the city looked impressive, that appearance was very deceptive. In 1636, an English traveller named

Witches Well, Castle Hill.

Sir William Brereton described the High Street as 'the most stately and graceful street that I ever saw in my life.' Then he added:

> Were not the inhabitants the most sluttish, nasty and slothful people … I never came to my own lodgings in Edinburgh or went out, but I was constrained to hold my nose.

The fact that residents simply emptied their chamber pots out of the windows into narrow streets didn't help.

Since they didn't have much else going for them, the Scots became more and more fanatical about their religion. They convinced themselves that they, not the Jews, were God's 'chosen people' and that their way of worship was the only one that counted – an attitude that would eventually destroy them. I'm sure there's a contemporary lesson to be learned there.

At first, James attempted to impose English Anglican practices in Scottish Kirks, but the Scots were having none of it so he wisely backed off. He died in 1625 and the throne passed to his son Charles I, a different kettle of fish entirely. The last British monarch to be born in Scotland, Charles had none of the political acumen or moral flexibility of his father and believed utterly in the divine right of kings. His policy was 'do what I say, or else' and that included the Scots.

The Covenanters

In 1637, Charles attempted to introduce an Anglican prayer book into Scotland at a sermon in St Giles', which he had 'promoted' to the status of Cathedral. According to legend, a woman called Jenny Geddes launched her stool at the Minister's head shouting, *'Dauir ye say Mass in my lug?'* (How dare you say Mass in my ear?) and a riot ensued. Insurgency, as we shall discover, was almost a pastime in Edinburgh.

This would have been a good moment for Charles to cut his losses. An arrogant, humourless and distant figure, devoid of his father's charm, he wasn't even popular in England. Instead, he refused to back down and what was a protest rapidly became a movement. Sound familiar?

As a Presbyterian minister, and one of the leading lights of the Scottish Reformation, Alexander Henderson was mortally offended by Charles' bull–headedness. So, after the debacle at St Giles', he helped draw up a document known as the *National Covenant*.

Its main premise was simple. The Scots were happy to keep Charles I as king, but only if he stopped mucking about with their religion. In 1638 a gathering of the clergy, nobles and common people ratified the Covenant in Greyfriars Kirkyard and copies

Alexander Henderson, Greyfriars.

were sent round the country for others to do likewise. One of the first 'Covenanters' to sign was James Graham, Marquess of Montrose, reputedly with his own blood. Appropriate, really, considering all the blood that was about to be shed. For two people who began on the same side, Henderson and Montrose would end up taking drastically different, if equally tragic, paths.

Henderson may have done too good a job on the Covenant. Despite the fact that he had lived happily in Leuchers for thirteen years, the General Assembly, governing body of the Scottish Kirk, promptly moved him to Edinburgh. He wasn't best pleased, partly because Charles had decided to play hard ball and marched north with an army of 20,000 men. When the Covenanters called his bluff and sent their own army to meet him, the king backed down. Not for long, though.

In 1640 he planned a new invasion, so the Covenanters marched south under Montrose and forced him to retreat again. Meanwhile, discontent with Charles was reaching breaking point in England, where the autocratic king would recall parliament each time he needed money for his campaigns, then dissolve it again if he didn't get exactly what he wanted. In the end, the 'Parliamentarians' got so sick of his shenanigans that they rebelled too, kicking off the English Civil War.

The English Civil War

The Parliamentarians asked for the help of the Covenanters, who picked the lesser of two evils and agreed to assist. Mainly because they saw an opportunity to advance their cause, a move which was to backfire spectacularly.

By this time, Henderson had risen to the positions of Moderator of the General Assembly and Rector at Edinburgh University, so was considered perfect for the task of furthering the Covenanters' aims. In 1643 he drew up the *Solemn League and Covenant*. In essence, it stated England would have to adopt Presbyterianism if they wanted the Scots to fight Charles.

In five short years the Covenanters had gone from wanting to be left alone to insisting everyone else worship like them. I'm sure there's a contemporary lesson to be learned there.

The Parliamentarians, also choosing the lesser of two evils, acquiesced. With the Covenanters' help, they defeated Charles, and the tangled web they had woven for each other became ever more snarled.

The story of the Covenanters is hideously complex but I feel it's necessary to give an accurate timeline of what happened, though it rather beggars belief.

Many Scots, seeing Charles was in danger of losing the throne entirely, balked at the idea of deposing their anointed king. The Marquis of Montrose was the first to switch sides. Leading a heavily outnumbered army of Catholic Irishmen and highlanders, he inflicted a series of defeats on the Covenanters, before fleeing for France.

Meanwhile Edinburgh was suffering the Great Plague of 1645, the last and worse of many outbreaks that had afflicted the city. It killed well over half the population of the district of Leith, Parliament fled to Stirling, grass grew on the Royal Mile and parents could be hung or drowned for concealing a sick child.

God seemed to have thumbed his nose at the chosen people, but this only hardened the Covenanters' resolve. Thinking they weren't being zealous enough, they began to turn on their most moderate factions and become even more radical.

Getting trounced by the Parliamentarians down south, Charles surrendered to the Presbyterians in the hope of persuading them to take his side. Yet he still insisted he wouldn't accept their terms, though Alexander Henderson is said to have begged him, weeping, on his knees. The disgruntled Covenanters handed their stubborn king right back to the Parliamentarians, now led by Oliver Cromwell, in exchange for £20,000.

Henderson, his health ruined by the strain of his fruitless negotiations, died shortly after and was buried in Greyfriars.

Cromwell considered the north paid off, until Charles flip flopped again. He did a secret deal with the more reasonable Covenanter factions, who invaded England to free him, re–igniting the English Civil War. See what I mean about complex? This time they were decisively defeated, leaving the extremists, led by the Marquis of Argyll, in charge of Scotland.

The Parliamentarians decided enough was enough. In 1649 they executed Charles I and declared England a republic, considering that an end to the matter.

The Scots were livid. Charles may have been an inflexible pain in the neck, but he was still a Stuart. The Covenanters immediately opened negotiations with his exiled son, the future Charles II, who reluctantly agreed with their demands to make England Presbyterian in return for their support. Charles even disowned the ever loyal Montrose, who had returned to fight for him. Without the backing of his king, Montrose was defeated, captured and executed – and his head placed on a spike outside St Giles'.

Still, he has a very impressive tomb inside St Giles' Cathedral, right next to a copy of the National Covenant.

Faced with the threat of another Stuart king, Cromwell marched north and occupied the city, the Flodden wall doing nothing to stop him. He turned the church of Greyfriars into barracks and set up cannon in St Cuthbert's. Seeing the writing on the wall, Charles II high tailed it back to France and Cromwell became 'Lord Protector of the Commonwealth', leaving Scotland under military occupation.

Tomb of Montrose, St Giles' Cathedral.

The Restoration of the Monarchy

In 1658, Oliver Cromwell died and his son, Richard, couldn't hold onto power. Cromwell's military governor in Scotland, General Monck, decided to act before England was plunged into chaos. He called a meeting of the Scottish clergy in Greyfriars, switched sides (there's a surprise) and marched on London. Heartily sick of being ruled by po–faced killjoys, the English backed down and invited Charles II to take over. He was finally crowned king of Britain in 1660.

Still mad at the Covenanters for trying to dictate to him, the new monarch had Argyll executed and his head put on the same spike Montrose once occupied. Funny how history tends to repeat itself.

Argyll now has a monument directly opposite Montrose in St Giles. When you walk in you almost feel the tension.

Charles II then declared Scotland to be Anglican and set about systematically destroying the exhausted Covenanters.

The Killing Time was about to begin.

Alexander Henderson's role in the Covenanter movement certainly wasn't forgotten by his enemies. If you look closely at his memorial you can see that it is one of the markers covered in tiny pockmarks. They are the indentations of musket shots, where English forces used his last resting place as target practice.

Now we move to one of the finest tombs in Greyfriars, belonging to a man who made an indelible mark on Edinburgh's history. He was called Sir George Mackenzie but has become better known by another name.

Bloody Mackenzie.

George Mackenzie was a remarkable man. He defended the Covenanter leader, Argyll, at his trial and argued against the persecution of witches, protesting that they were mainly harmless old women. He also founded the Advocates Library, which became the National Library of Scotland.

Yet, when Charles II appointed him Lord Advocate in 1677, Mackenzie set about persecuting the Covenanters with a zeal that was truly disturbing. Riven by internal strife and hunted by the authorities, Presbyterians were forced to hold secret outdoor 'Conventicles', preaching to dwindling congregations. They rose up in rebellion twice, until conclusively defeated at the Battle of Bothwell Brig.

Tomb of
Argyll, St Giles'
Cathedral.

Covenanter's
Prison,
Greyfriars.

After the battle, 1,200 captured rebels were marched to Edinburgh and locked in an area known as the Covenanter's Prison. Now incorporated into the Greyfriars Graveyard it was, in effect, one of the world's first concentration camps.

It was mid–winter and many perished. Some recanted and were set free but 400 who refused were eventually put on a ship bound for the West Indies to be sold into slavery. It sank in a storm off the coast of Scotland and, locked in the hold, the Covenanters drowned.

Seems like God wasn't on their side after all.

George Mackenzie mausoleum, Greyfriars.

The Glorious Revolution

The most ironic aspect of the Covenanter story is that southern Scotland became Presbyterian once more as they'd wanted all along. When Charles II died, his brother James became king, but was suspected of being pro–Catholic and pro–French, an untenable combination that saw him deposed after three years. In 1688, the English invited James' Dutch son–in–law, William of Orange, to take over, establishing the Hanoverian Dynasty. To appease the Scots, many of whom were still loyal to the Stuart line, William made Presbyterianism the official religion of Scotland. Not that Edinburgh's citizens were particularly placated, for they protested by destroying the Chapel Royal in Holyrood and ransacking its tombs.

As always, politics and loyalties proved fluid in such turbulent times, as evidenced by the fate of Reverend David Williamson. Though he had preached to Conventicles and was a Covenanter captain at Bothwell Brig, he was given back his old job as Minister of St Cuthbert's and eventually became Moderator of the General Assembly.

George Mackenzie, who had opposed the dethronement of James, also emerged unscathed. He wisely retired from public office and died peacefully in 1691.

Or did he? It is said he can never rest in peace after the atrocities he committed and his coffin is supposed to move around in its magnificent tomb. Perhaps he's just annoyed that Greyfriars is also the site of the Martyrs' Memorial.

Erected in the 1771, it commemorates the Covenanters hanged in the Grassmarket, over a hundred of whom are buried in a trench to the west of the monument. In total, it is thought there were around 18,000 Presbyterians executed in Scotland, many on Mackenzie's orders.

His legacy has a rather bizarre conclusion. The Covenanters' Prison is reputed to be the lair of 'The Mackenzie Poltergeist', the best documented supernatural case of all time. The 'City of the Dead' Ghost Tour, which has sole access to the prison, has recorded hundreds of attacks on their customers, including people bitten, cut, burned and knocked unconscious. I guess nobody likes a tourist.

David Williamson plaque, St Cuthbert's.

Martyrs' memorial. Greyfriars.

However, after more that 300 years of being rulers, the Stuarts (or Stewarts as they were now called) weren't going to slink off into exile. Instead they would plunge Britain into yet more conflict – with Edinburgh playing a major role.

So far I have concentrated on St Cuthbert's and Greyfriars, as they were the only prominent graveyards around at the time. Now another was founded.

Canongate Kirkyard.

Canongate Kirkyard

The graveyard surrounds Canongate Kirk on the lower Royal Mile and replaced the burial grounds of Holyrood Abbey and Lady Yester's Kirk. The church itself is considered unique among seventeent–century places of worship, a symmetrical cruciform with a distinctive curvilinear ashlar Dutch gable. In other words, it resembles a giant stone carriage clock.

The graveyard, on the other hand, is less impressive. Once looking up to the imposing sheer face of Calton Hill and surrounded by trees, it is now ringed by modern buildings of dubious architectural merit, unless you like discoloured harling. Still, it keeps the wind out.

As with all Edinburgh burial grounds, however, it has some great stories. This one is my favourite.

Canongate Kirkyard and Kirk.

Ebeneezer Scroggie, buried in 1836, was a city vintner and corn merchant. In 1841, a visiting Charles Dickens took a stroll through the churchyard and spotted his tomb, which read 'Ebeneezer Scroggie: Meal Man'. Dickens misread it as 'Mean Man' and it got him thinking. What kind of bloke was so stingy even his marker mentioned it? Two years later he published *A Christmas Carol* featuring the famous miser, Ebeneezer Scrooge.

The most Christmassy book of all time is not a religious tome. Ironically, it's a celebration of ordinary people, with turkeys and presents and dancing and everything that was disliked by the Presbyterian clergy. Christmas had always been a sombre, low key affair in Britain, a celebration of the birth of Christ, rather than an excuse to sit in front of the telly stuffing your face and watching re–runs of *The Great Escape*.

A Christmas Carol changed all that. Dickens was the most famous writer in the world and his books had a huge influence on the public. The massive success of his 'Ghostly Little Book' changed the way we thought of the festive season forever. From that moment Christmas began to morph from a strictly religious ritual into a secular holiday and remains so to this day. John Knox would be turning in his grave, if there wasn't a Honda parked on top.

Another irony is that Scroggie was the opposite of Ebeneezer Scrooge. The real man was generous, threw wild parties, reputedly fathered a child after having sex with a serving girl on a flat gravestone and was almost thrown out of a General Assembly debate for grabbing the bum of the Countess of Mansfield.

His marker is gone now, destroyed by construction work on the graveyard in 1932. I've done my best to pinpoint where it was and taken a picture of the approximate location. This may be unsatisfactory to some, as it's just a scrubby patch of grass, but that's the fickle hand of fate for you.

As a footnote, parents stopped calling their offspring Ebeneezer after the publication of *A Christmas Carol*, leaving Mr Scroggie as one of the last holders of such an impressive handle.

Approximate location of Ebeneezer Scroggie, Canongate.

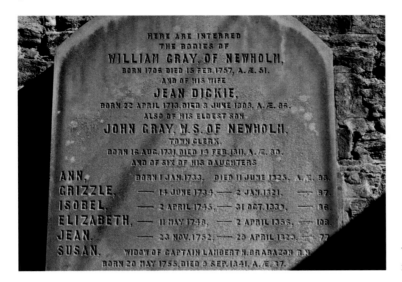

HERE ARE INTERRED
THE BODIES OF
WILLIAM GRAY, OF NEWHOLM,
BORN 1706 DIED 15 FEB.1757, A. Æ. 51.
AND OF HIS WIFE
JEAN DICKIE,
BORN 22 APRIL 1713, DIED 8 JUNE 1809, A. Æ. 96.
ALSO OF HIS ELDEST SON
JOHN GRAY W.S. OF NEWHOLM,
TOWN CLERK,
BORN 16 AUG.1731, DIED 19 FEB.1811, A. Æ. 80.
AND OF SIX OF HIS DAUGHTERS
ANN,   BORN 1 JAN.1733, DIED 11 JUNE 1825, A. Æ. 93.
GRIZZLE,   —— 14 JUNE 1734, —— 2 JAN.1321,   —— 87.
ISOBEL,   —— 2 APRIL 1745, —— 31 OCT.1833,   —— 86.
ELIZABETH, —— 11 MAY 1748, —— 2 APRIL 1855, —— 108.
JEAN,   —— 23 NOV.1752, —— 29 APRIL 1829, —— 77
SUSAN,   WIDOW OF CAPTAIN LAMBERT H. BRABAZON R.N.
          BORN 20 MAY 1755, DIED 3 SEP.1841, A. Æ. 17.

William Grey and family, Greyfriars.

This is a rather fun feature of old graveyards. They show us names that no longer exist. The eighteenth–century marker of William Grey in the Covenanters' Prison, for instance, also celebrates his daughters, Anne, Isobel, Elizabeth, Jean, Susan and Grizzle.

Grizzle? Seriously?

Grey's marker also demonstrates how large families were in the past (William had a son as well). Life expectancy was short and child mortality rates high, especially among the poor, so people compensated by having lots of kids. Another look at the marker, however, tells you the Greys were comfortably off. Not only could they afford a memorial on a wall but they all lived to ripe old ages. (Grizzle managed eighty–seven and Elizabeth an astonishing 108).

The Union of Parliaments

Years of war and failed crops had drained Scotland's economy so, in the late seventeenth century, Edinburgh's parliament staked everything on the 'Darien Scheme'. This was an attempt to set up Scotland's first international trading colony on the Isthmus of Panama. Thousands of people invested in what turned out to be a complete disaster, for the place was a plague infested swamp where nothing would grow. Of the sixteen ships sent to Panama, only one returned, carrying a handful of survivors. The country was bankrupted and desperately needed English money and commerce to survive. Despite the usual rioting in the streets of Edinburgh, MP's voted to dissolve their own parliament and form a union with England. In 1707, the two counties were officially joined.

Edinburgh had first lost its royal court and now its place as the seat of government. As a result of this combination, a high volume of the aristocracy deserted the capital and moved south. The evidence is the lack of graveyard markers dedicated to upper classes after 1707, combined with a rise of those from the tradesmen's classes, in particular those of smiths and brewers. It seems we were drowning our sorrows over the loss and the Scots' relationship with drink has been a thorny one ever since.

Since we were no longer the seat of royalty *or* power, Edinburgh found itself at a cultural crossroads. Typically, the Jekyll and Hyde city split into two factions. Some poets (known as Makars) looked to the past, writing and publishing in broad Scots and attempting to revive indigenous traditions they feared would fade away. Of these, the most famous was the poet and bookseller, Allan Ramsey, a former wig maker who is also credited with founding Britain's first circulating library. He is now remembered as the author of the play *The Gentle Shepherd*, as you can see by looking at his marker. Yet his efforts to collect lowland Scots poetry had a more immediate influence. Ramsay's *Tea Table Miscellany* of 1724 paved the way for dozens of similar books, including David Herd's *Ancient and Modern Scottish Songs* of 1776 and Walter Scott's *Minstrelsy of the Scottish Borders* in 1802. It was a movement the historian David Daiches called 'Nostalgic Patriotic Antiquarianism', which is a great way of saying 'good old fashioned Scots writing'.

Ramsey also fell foul of the church through his involvement in the theatre. Surprisingly, considering Edinburgh's present reputation as a festival city, theatre was looked on as the dominion of the devil and intermittently banned. Then again, if you've ever seen *Cats*, you might tend to agree.

Despite his fame, Ramsey was interred in an unmarked grave in Greyfriars and didn't get a proper monument until the nineteenth century. Like so many markers, it certainly isn't in the right place, unless Ramsay was buried half way up Greyfriars' Kirk wall.

Other citizens accepted the union and all the benefits it entailed. They were still patriotic but believed that Scotland, and Edinburgh in particular, could establish itself as a vital and appreciated part of Great Britain. They also felt that anyone who wanted an audience for their work should write in standard English.

Graveyard markers reflect this. Once inscribed in Latin, they slowly transformed into Scots, up until the union. Then gradually, English took over as the main language in honouring the dead, which is handy for me as I don't read Latin and most Edinburghers struggle with old Scots. So it's not hard to see which faction won this particular struggle.

In fact, the Unionists succeeded beyond their wildest dreams, for a culture was emerging in the capital that would change the world. The south had always been much richer than Scotland and the Darien disaster only widened the gap. Yet, due to John Knox and the Reformers, we had a general education system that was far superior to England and boasted five universities to their two. That difference would soon have a profound effect on the capital. Seeds planted back in the Reformation were beginning to take hold, and nobody better exemplifies this than the man who began Edinburgh's famous Botanical Gardens.

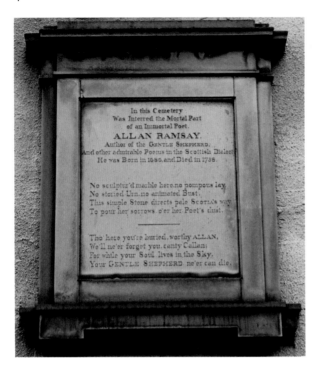

In this Cemetery
Was Interred the Mortal Part
of an Immortal Poet.
ALLAN RAMSAY.
Author of the GENTLE SHEPHERD,
And other admirable Poems in the Scottish Dialect
He was Born in 1686, and Died in 1758.

No sculptur'd marble here, no pompous lay,
No storied Urn, no animated Bust;
This simple Stone directs pale Scotia's way
To pour her sorrows o'er her Poet's dust.

Tho' here you're buried, worthy ALLAN,
We'll ne'er forget you, canty Callan;
For while your Soul lives in the Sky,
Your GENTLE SHEPHERD ne'er can die.

Allan Ramsey plaque, Greyfriars.

Robert Sibbald was the first professor of medicine at Edinburgh University and founded The Royal College of Physicians in 1681. His fame was so great, the Blue Whale was originally named *Balaenoptera Sibbaldus* after him, which impresses the hell out of me. Shame it was later changed to *Balaenoptera Musculus*. There's still a wild rose called *Sibbaldia Procumbens*, but that's a bit of a demotion from the largest mammal on the planet.

Never mind. By the end of the seventeenth century, Sibbald's efforts helped propel the College into one of the leading centres for science in Europe. Edinburgh's reputation was further enhanced in 1726 when Professor Charles Alston (buried in Canongate) and the six times Lord Provost, George Drummond, founded The Edinburgh Medical School.

However, the law of completely unintended consequences, which I just made up, meant these august bodies were to play a truly macabre role in later years.

On the positive side, the next graveyard we will explore was opened.

In 1718, Old Calton Cemetery replaced the now vanished South Leith Parish Burial Ground. Cut down in size when Regent Road was built through it, the cemetery now hides behind a giant wall blackened with age and traffic fumes. On the other side are mausoleums galore and moss–covered markers jutting out at all angles that wouldn't look out of place in some gothic vampire movie. Perched on a small hillock overlooking the Old Town, one side looks across a rooftop vista to the volcanic crag of Arthur's Seat. On the other side the monuments of Calton Hill frame the skyline. There are many truly remarkable men buried in there, but we have to give them time to die, so we'll come back to Old Calton soon.

Heart of Midlothian, St Giles'.

Bricks showing outline of the Tolbooth Prison, St Giles'.

secretly make wax imprints of the locks and keys. When the hauls he had gotten away with weren't large enough, he enlisted the help of accomplices on his largest escapade. Attempting to rob the Edinburgh Excise Office in 1788, he took with him George Smith, Andrew Ainslie and Humphry Moore.

The robbery was a fiasco, with Brodie reputedly falling asleep during its execution. That's what happens when you have two mistresses, I suppose. Humphry Moore turned Kings Evidence, Smith and Ainslie were arrested and Brodie fled to Holland, where he intended to book passage to America.

The Deacon proved again that he was no criminal mastermind. Posing as 'John Dixon' he gave a fellow traveller named Geddes a pile of goodbye letters to take back to Edinburgh, including one to his mistress. Geddes contacted the authorities and the Deacon was arrested in Amsterdam.

By this time Ainslie had also confessed and Smith and Brodie were hanged for the Excise Office robbery, on a scaffold Brodie himself had helped design.

Deacon Brodie may have been a cad in life but he faced death in a much more heroic manner. On the scaffold he chatted casually with his friends and, when the drop had to be delayed twice due to technical difficulties, he wittily remarked, 'After all, one has to get used to these new contraptions.'

Talk about gallows humour.

Deacon Brodie is often cited as the inspiration for Robert Louis Stevenson's *Dr Jekyll and Mr Hyde* and there's no doubting his influence, for the teenage Stevenson wrote a play about him. Yet Brodie was a bad guy pretending to be a good one, which isn't quite the same thing. I believe the more mature Stevenson was making parallels with Edinburgh itself and, by extension, all of mankind. In this, *Dr Jekyll and Mr Hyde* had far reaching consequences nobody could have anticipated. If I'm right, it affected the world in a way that is utterly appalling, as we shall see.

Law and Order

An ordinary man could reverse his fortunes to attain a position of wealth and respectability, but it took a lot of perseverance. The story of Peter Williamson, buried in Calton Cemetery, demonstrates this, and shows how brutal life in eighteenth century Scotland could be. Though his grave is unmarked, it is just north east of the Political Martyr's Monument.

Williamson's parents were poor crofters, so he was sent to live with an aunt in Aberdeen, where his incredible run of changing fortunes began. Playing on the quay in 1743, the eight year old was kidnapped and transported to Philadelphia to be an indentured servant for seven years – basically a white slave. This was common practise in those days and, remarkably, his abductors were aided by Aberdeen Baillies (civic officers). Peter's master in America, Hugh Wilson, was a Scot who had also been abducted and sold into slavery. This may explain why, when he died, Wilson left Williamson £120 in his will. With this fortune, Peter married and settled down to farm, but was captured by Cherokee Indians and made their slave instead. When he finally escaped, his wife was dead, so he joined the British army fighting the French – and was captured yet again. The guy was obviously a slow runner.

He was finally exchanged for a French prisoner and arrived, flat broke, in York. Determined to walk back to Scotland, his story aroused enough interest that he was persuaded to turn it into a book, affording him the funds to reach Aberdeen. He also took to dressing like a Native American, earning him the nickname 'Indian Peter'.

Approximate location where Peter Williamson
is Buried, Old Calton.

Aberdeen was not welcoming. Williamson was arrested by the same magistrates he had accused of kidnap, sued for libel and made to sign a statement stating the story was a lie. After that, he was fined and banished.

Turning up in Edinburgh, he ran a coffee house frequented by lawyers, who helped him sue the Aberdeen magistrates. Given the dislike between the lowlands and highlands, it was no surprise that he won and was awarded a substantial sum. With the money, he opened a tavern and printing shop, inventing his own portable press and waterproof ink. He launched a weekly magazine, compiled Edinburgh's original street directory and founded the first regular postal service in Britain. Perhaps fittingly for a man called Indian Peter, who owned a tavern, he became an alcoholic and died in 1799 – probably of liver disease.

It was an end met by many Edinburgh residents, who were all serious drinkers, usually starting at breakfast time and continuing sporadically until 10.00 p.m. Even respected lawyers and judges would conduct their business inebriated or with monumental hangovers.

I'm particularly fond of the judges of this period, who could easily be classed as villains in any other city. Scotland's system of law was considered one of the most advanced in the world, but you wouldn't think so by looking at its practitioners. Drunken Lord Braxfield, who sentenced Deacon Brodie, was so gung–ho he was called 'The Hanging Judge'. When a political reformer being tried for sedition complained that Jesus was also a reformer, Braxfield countered, 'Muckle [much] he made o' that. He was hangit.' Another choice phrase was 'Hang a thief when he's young and he'll no steal when he's old'.

Or we have the, equally flippant, Lord Kames who once sentenced a chess–playing companion to death with the quip, 'Checkmate, Matthew.' An agricultural improver, moral philosopher, literary critic and historian, he was one who still relished the coarseness of the Scot's language, as did Braxfield. On his last visit to court, aged eighty–seven, Kames bid farewell to his legal brethren with a cheery 'Fare ye a' weel, ye bitches!' making him sound like the world's first rapper.

The New Town

Edinburgh's fortunes seemed to be finally changing. Lord Provost George Drummond was determined it should expand out of the dirty, smelly, brutal, booze sodden Old Town and he was visionary enough to pull it off.

Drummond is a fascinating character, who wasn't afraid of getting his hands dirty. As well as co–founding the Edinburgh School of Medicine, he founded Edinburgh Royal Infirmary, helped start the Royal Bank of Scotland and instigated the draining of the Nor Loch to make Princes Street Gardens. He fought against the Jacobites at the battle of Sherrifmuir in 1715 and was one of the few who tried to defend Edinburgh from Bonnie Prince Charlie. When he died the *Scots Magazine* published this obituary.

'He engaged in public business at the age of eighteen, and continued capable of discharging it till the end of his life. During that long period, scarce any scheme for the improvement or advantage of this country has been carried on, of which he was not an active and able promoter.'

George Drummond, Canongate.

Today's civil servants should hang their heads in the face of his achievements.

In 1766, he put his plan for expansion into action by organising a competition to design a New Town in the north. It was won by twenty–seven–year–old James Craig.

The final design was deceptively simple, consisting of three parallel streets (Princes Street, George Street and Queen Street) with a square at each end (St Andrews and Charlotte squares). Legend has it the layout was popular because it allowed the wind to whistle down each thoroughfare, carrying away bad smells. Whatever the reason, it is now considered a masterpiece. Along with the Old Town, it was declared a UNESCO World Heritage site in 1995, albeit an exceedingly breezy one.

Surprisingly, Craig subsequently struggled to find work and ended up buried in Greyfriars, back in the middle of the Old Town. Again, I can only hope I've got the right place. Though Craig died in 1795, the complex ban on memorials in overcrowded Greyfriars meant his flat marker was only laid in the 1930s.

The final development of Charlotte Square was taken over by Robert Adams, of the far more respected and famous Adams' family of architects. If you look at the picture of their mausoleum in Greyfriars you're left in no doubt you've got the right spot this time.

I've lived in flats that were smaller.

Now the richer citizens of Edinburgh had somewhere nice to move and did so in droves. Perhaps the more aesthetic surroundings inspired them, for the city was about to enter a golden age as the centre of the Scottish Enlightenment. In fact it was so pivotal that many refer to it as the 'Edinburgh Enlightenment'.

Adams' family mausoleum, Greyfriars.

James Craig,
Greyfriars.

The Enlightenment

Up to this point, Edinburgh had been a major influence on Scotland and England. Even North America, if you consider the impact fleeing Covenanters had in establishing Presbyterianism in what is now the USA.

This time, we were about to change the entire world, leading the planet in medicine, philosophy, law, economics, science and architecture. This renaissance was markedly different from the rest of Europe, which had gotten all arty. Due to the effect of the Reformers who, ironically, would have hated it – the Edinburgh Enlightenment was heavily science based. It stressed the fundamental importance of human reason, combined with a belief in the ability of that reason to effect changes for the better in society. Easier said than done, but it was a start.

Edinburgh soon earned the nickname 'Athens of the North', a phrase reputedly coined by the painter Hugh William Williams, who is buried in Canongate.

If I were to include every Edinburgh marker and monument celebrating the heroes of the Enlightenment, I'd end up with something resembling a telephone directory. So I'll stick to the big guns, and that's a powerful enough armoury.

To my mind, David Hume easily ranks with the greatest thinkers of all time. Perhaps the greatest. As a philosophy graduate, I would love to give an in–depth critique of his ideas, but we'd be back with a telephone directory–sized book.

To put it simply, Hume espoused 'Empiricism', the idea that knowledge comes primarily from sensory experience – placing ultimate importance on the evidence of

Hugh William Williams,
Canongate.

experimentation. Note to all Quantum Physicists. You're eventually going to come round to his way of thinking again. He also insisted that morals were subjective, not objective, which is true, no matter how unpalatable you might find the idea.

Many people did find Hume unpalatable. When he died in 1776, the good old Edinburgh Mob struck again, probably with a dead cat in each hand. His grave had to be guarded for eight days after burial, largely due to his professed atheism, not a popular ideology after the Reformation and Covenanters. Fortunately, Hume's magnificent mausoleum is still intact. It was just too darned big to knock down.

We also have Adam Smith, 'Father of Economics' and author of *An Inquiry into the Nature and Causes of the Wealth of Nations*. He espoused free market capitalism, a term he made famous, and which is still practised as the financial model in most developed nations. It is said Margaret Thatcher carried a copy around in her handbag, though we can't really blame Smith for that.

Then there's James Hutton, 'Founder of Modern Geology'. A huge influence on Charles Darwin, his works disproved the biblical interpretation of how the earth began, supplanting it with the idea that the landscape was formed over eons. He is remembered by a miserable little plaque in the Covenanters' Prison, when he should have a monument twenty feet high.

In the poetry corner is Robert Fergusson, the man Robert Burns considered his literary superior, who died at the age of twenty–four in 1774. Injuring himself after falling down a flight of stairs and suffering from delirium, he was incarcerated in the Edinburgh Bedlam asylum, just outside Greyfriars, where he rapidly succumbed to its horrors. He wrote 'Auld Reekie' (Old Smokey), a vivid poem about his home town and popularised the Standard Habbie, a form of verse that Burns made famous. In acknowledgement, Burns designed and paid for Fergusson's marker in Canongate. It's a measure of the poet's standing that Robert Louis Stevenson planned to add a stone epitaph, but died before he could do so. However, Fergusson has a statue right outside Canongate, erected in 2004.

Left: David Hume mausoleum, Old Calton.

Right: Adam Smith, Canongate.

Below: Hutton plaque, Greyfriars.

Other notables are John Playfair, scientist and mathematician, who backed up Hutton's work and Joseph Black, physician and chemist, who discovered carbon dioxide and latent heat (leading to the science of Thermodynamics and the development of the steam engine). Playfair has an insignificant plaque over his burial place in Old Calton, so I have opted for a picture of his imposing memorial on Calton Hill next door.

Architecture has the previously mentioned Adams family, who influenced building styles in Western Europe, North America and Russia. Art has the painters Allan Ramsey, son of the poet and publisher and the Runciman brothers.

Music is represented by the noted composer and fiddler, Daniel Dow, and the operatic composer and bassoon player, John Fredrick Lampe, both buried in Canongate. Remember, these are just the innovators who ended up interred in Edinburgh. I haven't included all the intellects whose final resting places lie somewhere else.

Less Recognised

The unfortunately named Thomas Bonar and William Smellie are lesser known than most Enlightenment innovators, but their product is world famous. Bonar was the co–founder and Smellie editor of the *Encyclopaedia Britannica*. For younger readers, that's how the people used to find out stuff before Wikipedia came along.

You may have noticed that we are up to the mid–eighteenth century but none of the famous markers I've covered so far belongs to a woman (Phoebe Traquair doesn't count as she died in the twentieth century). Canongate does have Agnes Maclehose (Clarinda) and Euphemia Murray (The Flower of Strathmore), but their claim to fame is that they both caught the eye of a womanising Robert Burns. As you can see from the picture, time hasn't treated Euphemia's grave with much respect either.

Why no females in the arts and sciences? A glance at the original *Encyclopaedia Britannica*, gives us a disturbing insight. Smellie's entry for 'Woman' simply states, 'the female of man', which is about as dismissive as you can get.

Eighteenth–century Edinburgh wasn't a great time to be a member of the fairer sex either.

That speaks volumes to me. God knows how many erudite and accomplished females haven't had the recognition they deserve, in life or death. Things do change as the years roll on, but women are vastly under–represented in older graveyards.

My favourite character of the Enlightenment didn't get the recognition due to him either. He has no plaque or marker but is interred in the Covenanter's Prison, in the Mausoleum of Patrick Grant of Elchies. Hope Patrick doesn't mind.

Left: Fergusson statue, Canongate.

Right: Playfair memorial, Calton Hill.

Below: Fergusson's grave, Canongate.

Left: Joseph Black, Greyfriars.

Right: Runciman Brothers, Canongate.

Below: William Smellie, Greyfriars.

Left: Agnes Maclehose, Canongate.

Right: Euphemia Murray, Canongate.

Below: Thomas Bonar, St Cuthbert's.

Lord Monboddo,
Greyfriars.

His name was James Burnett, also known as Lord Monboddo.

When Charles Darwin suggested that man descended from lower life forms rather than being a divine creation, the idea rocked the foundations of society and threw established religion into frenzy. Nobody wanted to be a monkey's uncle.

That's nothing. Almost a hundred years before Darwin's bombshell, I believe Monboddo started the great naturalist on the path that led to his everlasting fame.

James Burnett was born in 1714 and worked his way up to become one of the most eminent judges at the Court of Session. Eccentric doesn't begin to describe him. He loved classical culture (he was fluent in Greek) and, though it had become all the rage to travel in Sedan Chairs, Greeks and Romans didn't do it, so neither would Monboddo. He chose to ride on horseback, no matter how bitter the weather, but did deign to put his powdered wig in a Sedan chair if it rained.

In 1778 he was visiting the King's Court in London, when there was an alarm about the safety of the ceiling. The entire chamber rushed outside with the exception of Monboddo, who sat where he was until the all clear was given and everyone trooped sheepishly back. When asked why he had not evacuated with the others, Monboddo explained he thought it was 'an annual ceremony, with which, as an alien, he had nothing to do'.

He had the kind of mind, however, which could easily accept that astonishing things might be true and it opened him up to all sorts of ridicule.

One of these was the idea of human evolution.

Almost a century before Darwin's *On the Origin of the Species*, Burnett wrote *Of the Origin and Progress of Language*. In it he stated that we evolved from lower orders, were related to apes, and that modern man originated somewhere warm like Africa, before spreading over the rest of the globe. That we learned to use simple tools, then form tribes, and then acquire speech. As society improved, Monboddo claimed, language evolved and physical prowess declined, leading to the development of the modern humans we see today. He predicted the Xbox!

Monboddo also wrote that man's unchecked intellect would ultimately bring about his destruction. I'm sure there's a contemporary lesson to be learned there.

Did Charles Darwin know of his work? His grandfather Erasmus Darwin wrote about Monboddo and Charles studied medicine at Edinburgh University, close to Burnett's burial place. Burnett's wild theories had made him infamous, so there is little doubt Darwin heard stories of the eccentric judge and his oddball notions. This one certainly, seems to have stuck in his mind. After all, that's how natural selection works.

A successor to Burnett at the High Court, Charles Neaves, certainly thought so, for he penned this ditty:

> Though Darwin now proclaims the law
> And spreads it far abroad, O!
> The man that first the secret saw
> Was honest old Monboddo.

Post Enlightenment

The influence of the Edinburgh Enlightenment simply cannot be underestimated. In 1750, the King's Chemist, Mr Amyat visited the city and famously remarked,

> Here I stand at what is called the Cross of Edinburgh, and can, in a few minutes, take fifty men of genius and learning by the hand.

I presume he didn't try it, or he would have certainly gotten some funny looks.

Though the Enlightenment faded towards the end of the eighteenth century, it spawned a whole new generation of equally notable Scots, including the philosopher James Ferrier, who coined the term epistemology (the study of knowledge).

Then there's Thomas Carlyle (philosopher), Walter Scott (writer) and George Meikle Kemp, the architect who designed the Scott Monument (largest to any writer in the world). He was buried in St Cuthbert's after drowning in the Union Canal, which is no mean feat, as it's only a few feet deep. It also somewhat detracts from his status as a genius. We have the painters John Watson Gordon, David Allan, and the Naysmith Family, Alexander and his son Patrick. The other son, James, wasn't to be outdone, becoming a famous engineer and inventor of the steam hammer.

Other scientists include James Watt (pioneer of the steam engine), Lord Kelvin (pioneer of Thermodynamics), William Murdoch (inventor of gas lighting) and James Clarke Maxwell, whose work Einstein described as 'the most profound and the most fruitful that physics has experienced since the time of Newton'.

This is not to say that Edinburgh had gotten all civilised, for its strange dichotomy

Left: David Allan, Old Calton.

Right: Naysmith family, St Cuthbert's.

Below: James Ferrier, St Cuthbert's.

John Watson Gordon,
Canongate.

continued. In 1793 Thomas Moore and four accomplices, influenced by the French Revolution, campaigned for common man's voting rights. They were charged with treason and sentenced to fourteen years at Botany Bay in Australia. The judge? Lord Braxfield, of course. Their efforts are now commemorated by the gigantic Political Martyr's Monument in Old Calton.

The next year, none other than Walter Scott was involved in a pitched battle with Irish students at the Edinburgh Theatre, when they booed the national anthem. Despite its veneer of sophistication, this was still an incredibly volatile and base city.

Speaking of Walter Scott, he was instrumental in Edinburgh's next great transformation, one that changed Scotland forever.

The Kings' Jaunt

Hanoverian George IV was not popular. In 1822, to keep him from meddling in international affairs, his ministers suggested a trip to Scotland, a country which hadn't had an official visit by a British monarch for two hundred years. Knowing nothing of the animosity between north and south, the king decided wearing highland dress would be appropriate. Having no real idea what normal highland outfits looked like, George had one designed, paying an amount for it that most clansmen could have lived on for several years.

Political Martyrs' monument, Old Calton.

The finished mockery was made of satin, velvet and cashmere in eye–popping scarlet weave, now sold all over Scotland as the Royal Stewart tartan. It was an affront to the simple belted plaid worn by the real highlander, but nobody was going to point that out to the monarch. The Scottish gentry wanted to make everything perfect for this great honour, and there was only one man thought equal to the task of organising such an immense event. Sir Walter Scott.

Scott had a truly astounding talent for mythmaking. A gentleman's club called 'The Company of Archers', for instance, suddenly found they were the 'ancient bodyguards of the kings of Scotland' – which was news to them. In the spirit of the occasion they turned up sporting tight–fitting Lincoln green and white satin. Nobody blinked an eye.

It became obvious that national garb of some variety was going to be the order of the day and everyone wanted in on the act. Unfortunately, the lowland Scots didn't have one and, despite what most people think, neither did the north. Clans were normally identified by their cap badges, for a rugged people who battled the elements daily didn't have the time or inclination to weave complex tartans.

Undaunted, the novelist dashed out a quick booklet titled *Hints addressed to the Inhabitants of Edinburgh and others in Prospect of His Majesty's Visit* and stated 'no gentleman is to be allowed to appear in anything but the ancient Highland costume'.

Edinburgh gentlemen weren't going to miss out on the shindig of a lifetime because they didn't have the right togs. Perhaps acting out the role of ancient highland warrior wouldn't be too bad if they could spruce up that dull northern garb to their own specifications, so the gentry set about designing brightly coloured costumes with gusto. So great was the demand that, in one week, three hundred unemployed tailors were hired in Edinburgh alone.

Outdoing them all, portly George attended a ball squeezed into a mini kilt, dripping with jewellery and sporting fleshy coloured tights underneath.

There was one last issue. The monarch expected to see some authentic northern warriors on his visit. Since George had no intention of going near the highlands, Walter Scott sent out a call inviting clan chiefs to Edinburgh. The novelist's reputation ensured that his request was not treated lightly. After all, Scott's writing had painted them in glowing colours and they owed him one.

What would real clans wear for this magnificent occasion? Highland dress, whatever vague thing that might be, had been banned for eighty years. Once again Scott came to the rescue. At his suggestion, chiefs ordered wholesale costumes from weaving firms, who were already making kilts for the army and had no problem in conjuring up fake 'traditional' clan tartans.

Decked out in their new, mass manufactured finery, the highlanders exhibited enough sartorial elegance to be allowed to parade in front of their Hanoverian master – head of a dynasty who had ordered the annihilation of their grandfathers.

It didn't matter. In the end, the visit was far more important to Scotland than it was to the king. Walter Scott had done more than put on a pageant. He had altered a nation's perception of itself. The highlanders parading through the heart of Edinburgh (not surprisingly in the circumstances) looked far more romantic than anyone imagined. The lowlanders, instead of being afraid and suspicious, found they were actually proud of their northern counterparts.

In 1822, for the first time in its history, Scotland became an undivided nation.

Walter Scott exceeded his wildest ambitions. This tartan pageant had made his totally bogus vision of Scotland a reality, one that has grown stronger with each passing year. As the historian John Prebble so accurately put it,

> No other nation has cherished so absurd an image, and none perhaps would accept it while knowing it to be a lie.

In a way, it was a wonderful lie, for it finally integrated all of Scotland. What Edinburgh didn't realise, however, was 'the king's jaunt' would result in it giving away a huge chunk of its own identity.

The connection to our graveyards? The wonderful Ebeneezer Scroggie did the catering for the whole affair. A tenuous link, I'll admit, but it's too good a story to miss out.

New Calton
cemetery, With
Holyrood Palace in
the background.

New Calton and St John's

Speaking of graveyards. As we reach the early nineteenth century, two new sites are founded, though they are often confused with existing ones – since they are right next to them.

New Calton also has a magnificent view, overlooking Holyrood Palace and the new Scottish Parliament buildings, with the Firth of Forth glimpsed in the distance. Seems a bit of a shame that these sites are filled with people who can't actually appreciate the scenery. The first graves were the numerous bodies disinterred by the building of Waterloo place and Regent Street over Old Calton and are now found in the northern part of the site. In 1818, the Society of Jews bought a section of the burial ground close to the southern wall, near ground reserved for unclaimed persons who had died in the Royal Infirmary or in the Calton Gaol. Though one of the most popular resting places for the poor, New Calton earned the impressive nickname 'Cemetery of Admirals,' as there are six of them buried there.

It was also the first cemetery to have a formalized layout from the beginning though, to my eye, it doesn't look all that different to St Cuthbert's or Canongate.

St John's opened right next to St Cuthbert's and it's difficult to see the join, so most visitors think it's the same graveyard. (St John's is a bit higher up and closer to Princes Street.) It too has a list of notable burials at the end of the book.

It's a good job that new cemeteries were being established for the old ones were full to overflowing. In 1779, Hugh Arnott gave this description of Greyfriars:

St John's graveyard
and church.

The graves are so crowded on one another that the sextons frequently cannot avoid in opening a ripe grave, encroaching on one not fit to be touched! The whole presents a scene equally nauseous and unwholesome. How soon this spot will be so surcharged with animal juices and oils, that, becoming one mass of corruption, its noxious steams will burst forth with the prey of a pestilence, we shall not pretend to determine; but we will venture to say, the effects of this burying ground would ere now have been severely felt, were it not that, besides the coldness of the climate, they have been checked by the acidity of the coal smoke and the height of the winds, which in the neighbourhood of Edinburgh blow with extraordinary violence.

It is possible that 200,000 people are buried in tiny Greyfriars, most without markers, turning what was once a valley into a mound of thinly covered corpses.

Crime

Queen Victoria was crowned in 1837, ushering in the Golden Age of Empire. With the rise of Great Britain as the most powerful entity on the planet, surely all of Edinburgh finally shared in the prosperity?

Not in the slightest. Moving into the nineteenth century, we find a large part of the city worse off than it had ever been.

With the exodus of the rich to the fancy New Town and the demolishment of the Flodden Wall, the Old Town should have finally had space to breathe. Sadly, it wasn't to be, for a whole series of events would now have a calamitous effect on Edinburgh.

The Agricultural Revolution, begun in the seventeenth century, reached an apex in the nineteenth century, displacing thousands of people from their land. At the same time, the 'Highland Clearances' were taking place, as landowners realized it was more profitable to have sheep on their property than people, resulting in the eviction of whole populations. The Irish potato famine forced masses to leave their home country and, if they didn't fancy the perils of emigrating to America, Scotland was the closest place to escape starvation.

Since there was no work to be found in the countryside and we only had four cities (Edinburgh, Glasgow, Dundee and Aberdeen), a large proportion ended up in the capital. This influx was almost entirely made up of the poor and displaced, so they were crammed into the Old Town, where the population doubled in the early years of the nineteenth century. The once imposing area became a dilapidated slum with titanic mortality rates, virtually segregated from the rest of the city.

Worse than that, the jobs in factories, mills and mines was backbreaking, the hours unbearable and the wages laughable. Not that anyone was laughing.

So, what were the choices if you didn't want to spend your life hunched over a machine for sixteen hours a day breathing in noxious fumes for a pittance?

One option was obvious. Crime.

Lawbreaking was endemic in Victorian Edinburgh. Since executed criminals were buried in unmarked graves, it's a little hard to track the individuals down. Yet evidence of their presence is there. Greyfriars graveyards, for instance, still has well preserved Mortsages.

These were designed to thwart the most unique set of criminals in Edinburgh. The body snatchers or Resurrection Men.

Remember those medical schools? They required fresh corpses to practise on and the only legal bodies they could get their hands on were those of criminals who had been condemmed to death. Since Edinburgh still led the world in anatomy, the lecturers needed a lot more stiffs than were available. So they paid under the dissection table for bodies stolen from graves, no questions asked. Mortis Safes were used to house the recently departed until they were too badly decomposed to be useful. Or, if you were poor, you simply sat on your relative's grave for a few days and nights. To prevent the body snatchers, New Calton had a watchtower built in 1820 and St Cuthbert's followed suit in 1827, as well as raising its walls to eight feet high.

James Hunter's tomb and the Coach Driver's Monument in Canongate Graveyard were both damaged by crossfire, when watchmen defended the Kirkyard from a raid by Resurrectionists. The marks of shots fired were noted as still being visible on the south side of this monument during a survey of the cemetery in the 1960s.

There was another unintended consequence of being at the forefront of medical research. The Anatomy departments of Glasgow and Edinburgh had become fascinated by the concept of 'Medical Electricity', and the notion that it could be used to bring the dead back to life.

Mortsage, Greyfriars.

New Calton watchtower.

At that time, Edinburgh Medical College was in Infirmary Street, close to Greyfriars, and one of the graduates was a young man named John Poldari. A few years later, Poldari found himself in Geneva with the famous poets Lord Byron and Percy Shelly and got pally with Shelly's teenage wife. One night Byron came up with a macabre idea. They would each write a horror story for the others to read and see which was the scariest.

Despite their fame as writers, Byron's and Shelly's efforts are forgotten. John Poldari, however, penned a short story called *The Vampyr*. Published in 1918 it was the first published vampire tale and featured a dentally challenged aristocrat based on Byron. It also heavily influenced Bram Stoker's *Dracula*.

Shelly's wife, Mary, had obviously been paying attention to Polidori's stories about his time near Greyfriars. In 1818, she published a worldwide bestseller about a monster assembled from stolen body parts and brought to life when a doctor passes electricity through it. She called her novel *Frankenstein*.

It was inevitable, I suppose, that someone would figure out an easier way of providing Edinburgh's anatomists with fresh bodies, rather than scrabbling about in the dirt at two in the morning, while looking over their shoulders for signs of the law.

In 1828, two Irish labourers named William Burke and William Hare decided not to wait for natural expirations. Instead, they murdered sixteen people over a period of ten months, and then sold their bodies to the respected anatomist, Dr Robert Knox. To ensure the corpses were unmarked, Burke would suffocate them by sitting on top of the victim, one hand over covering the mouth and the other stuffing fingers up their nose – a method of killing known as 'Burking'. Burke was hanged in 1829 after Hare turned King's Evidence , with his body, naturally, handed over the anatomists for dissection.

Though body snatching was seen by many (especially the anatomists) as a necessary evil, the outcry over Burke and Hare ushered in the Anatomy Act of 1832. This allowed people to donate bodies for research and effectively ended the practice of having to steal them. It also made the newly built St Cuthbert's and New Calton watchtowers completely redundant, so they were rented out as accommodation. The New Calton watchtower was occupied until 1950 and despite being only five metres in diameter once held a family of ten. No space went to waste in overcrowded Edinburgh.

The graveyards were also suffering the effects of a vastly increased population. In 1863 St Cuthbert's was declared full and the church subsequently fined for refusing to stop holding funerals there. Burials were an important source of revenue for the Kirk and this is Scotland, after all. Greyfriars, Old Calton and Canongate had also reached saturation point, resulting in the commission of a raft of cemeteries, much larger than the existing graveyards.

Warriston, Dean, Rosebank, Grange and Newington

In 1843 Warriston Cemetery was established. Designed by Edinburgh architect David Cousin, it was the city's first 'garden cemetery' and extremely pretty before it fell into disrepair.

The site has been repeatedly vandalised and parts are so overgrown that many of the markers are barely visible. Though restoration work has begun, neglect has actually had a positive effect. Warriston, which contains mature Dutch and Guernsey Elms, has been designated a Local Nature Conservation Site and the former rail line running through it is now a walkway. It was also the first cemetery in Edinburgh to

Warriston cemetery.

Warriston crematorium.

James Young Simpson,
Warriston.

Dean cemetery.

have a crematorium, opened in 1929, which is no fun at all for grave spotters.

Warriston's most famous grave is that of James Young Simpson, who carried on the city's scientific tradition by discovering the anaesthetic properties of chloroform.

Dean cemetery, built in 1846, was also designed by David Cousin and lies just above Dean village, next to the Water of Leith. It was reserved for the mainly middle and upper classes and holds many later Enlightenment innovators, like the pioneer of early photography, David Octavius Hill. Who says death is the Great Leveller? What other burial ground includes an area called 'Lord's Row'?

It certainly is a beautiful place. The curvilinear central layout with later rectilinear extensions is studded with trees, planted over a century ago and now fully mature. The fact that it is owned and preserved by the Dean Cemetery Trust means it looks much the same today as it did when it was opened and it is considered a valuable resource of Victorian art. There are markers and monuments designed by such august artists, sculptors and architects as Sir John Steell, William Brodie, John Hutchison, Francis John Williamson, Pilkington Jackson, Amelia Robertson Hill, the Rhind family, William Grant Stevenson, Henry Snell Gamley, Charles McBride, George Frampton and Stewart McGlashan.

The same year, Rosebank Cemetery was established, yet again designed by David Cousin. I'm betting this guy wore black a lot.

Rosebank is located at the junction of Pilrig and Broughton Streets and was extended in 1880. Its most poignant marker commemorates the Quintinshil Disaster of 1915, when 216 men of the Royal Scots died in a train crash on their way to board a ship for Gallipoli. Homeless and stillborn children are also interred here, the latter with a granite memorial bearing the inscription 'to all those children never known but always loved'. Because of Rosebank's proximity to the Port of Leith, it has numerous maritime links and a number of markers reading 'drowned' or 'lost at sea'.

A year later, in 1847, Grange Cemetery was founded. Guess who designed it? Wrong! It was the architect, David Bryce. Maybe David Cousins was on holiday. By coincidence, David Bryce once lived in the New Calton watchtower, which gives a

Rosebank cemetery.

whole new meaning to taking your work home. He also renovated Colinton Church and built its new tower. Even that quaint graveyard wasn't immune to expansion, with a cemetery added on in the late eighteenth century. If you'll forgive me, I've missed out the notable burials in Grange at the end of the book, as they are so similar in professions, dates and sheer numbers to Dean.

Newington Cemetery, established in 1848, ended Bryce's brief reign. Once again, it was designed by David Cousin, who was probably transported around in a coffin by this time. Cousin's name is even inscribed on the family headstone in Dean Cemetery, though he's actually buried in the USA. He just can't get away from the place.

Newington is in a sorry state too. It was vandalised many times and much of the cemetery is so overgrown as to be impenetrable. Like Warriston, however, it has become a haven for animals, so it's appropriate that Sir Charles Gibson Connell, former president of the Scottish Wildlife Trust, is buried there.

If you look at the markers in all these new cemeteries, you can chart another aspect of Edinburgh's progress. Memento Mori, in their famous form, began to vanish. Symbols of mortality, immortality and trade emblems ceased to be used and individual masons carving unique markers faded from prominence. The development of mechanical means of stone cutting now resulted in the establishment of funeral firms working from design books. Victorians wanting to show their prosperity and erudition began asking for Egyptian, Greek and Gothic style models.

Sandstone gave way to granite or white marble, imported from Italy. While white marble looks very sophisticated, it is rather fragile and many of the monuments are now crumbling. Not so erudite, after all, those rich Victorians.

Egyptian–style Victorian
tomb, Grange.

Later Nineteenth Century

In 1843, Edinburgh went through the last of its great historical disruptions called, appropriately, 'The Disruption'. As had happened with the Catholic Church, 300 years earlier, dissatisfaction was growing over the role of the Church of Scotland. Many thought they had drifted from the ethos of the Reformers, concerned with respectability, conformity and power and tied to the state, rather than dictating their own affairs. Things came to a head at the General Assembly, when one third of the ministers and elders walked out and established the Free Church of Scotland, picking as their first Moderator Thomas Chalmers.

This may not seem like a big deal in this material age. However, the Church of Scotland was never as powerful again, which may have led to another Jekyll and Hyde aspect of this city. Despite a rabidly Christian past, Edinburgh is an extremely secular city, with a surprisingly large Pagan population.

The 'Wee Frees' as Free Church members are affectionately called, have little influence in Edinburgh but exert a far more powerful force in the north and west of Scotland. A few years ago I was talking to a woman in Stornaway, who complained that she couldn't hang her washing out on Sundays, as it was a day of rest. If she tried, her Wee Free mother in law would take it down again. John Knox would have heartily approved.

For the common man and woman in Edinburgh there were more pressing problems than choosing sides in church wars. As evidence of that we'll head for the grave of poet Alexander Smith, buried in Warriston. He painted a vivid picture in words of how horrible the nineteenth–century Old Town still was.

> The Cowgate is in the Irish part of the City ... The inhabitants are morally and
> geographically of the lower orders. They keep to their own quarters and seldom come

Left: Alexander Smith, Warriston.

Right: Henry Littlejohn, Dean.

Below: Thomas Chalmers, Grange.

up to the light of day. Many an Edinburgh man has never set his foot in the street: the condition of the inhabitants is as little known to respectable Edinburgh as are the habits of moles, earthworms and the mining population.

Not a particularly empathetic bloke, then. If he tweeted that out these days he'd be trolled until he made a public apology.

Contrast this with Dr Henry Littlejohn, buried in Dean Cemetery.

Littlejohn was cited by Edinburgh's Arthur Conan Doyle as one of the inspirations for his famous creation Sherlock Holmes. He is also known as 'The Man Who Cleansed Edinburgh', which makes him sound like a war criminal. In fact, he was Police Surgeon for Edinburgh, Medical Officer for Health, Head of Edinburgh University's Chair of Forensic Medicine, lecturer in medical jurisprudence at the Royal College of Surgeons of Edinburgh and Commissioner of the Board of Supervision, which oversaw the city's sanitary conditions. This was definitely a man who ate lunch at his desk.

Unlike Smith, he sympathised with the plight of the working class. In 1865, he published his devastating *Report on the Sanitary Conditions of the City of Edinburgh*, which outlined depravation and poverty in the slums. His description of the Old Town is far less judgemental than Smith's but equally disturbing.

> Like the Canongate, and for the same reasons, the scene in the High Street, Lawnmarket and adjoining West Bow in the last half of the nineteenth–century was one not only of dirt, near starvation and chronic poverty, but almost incredible overcrowding. Thus in 1865 there were 646 people to the acre in the Tron area of the High Street.

It was also Littlejohn who declared the cemeteries of Old Calton, St Cuthbert's, Canongate and Greyfriars to be chock full and insisted nobody but existing families be interred there. A good job too or, by now, we'd need a ladder to get into them.

Appalled by Littlejohn's findings, the town council undertook a major renovation of the decrepit Old Town, transforming it into the far more pleasant area tourists see today. Yes, it's not as old as most people imagine. Between 1860 and 1900, almost two thirds of the ancient buildings in the Old Town were demolished.

Out of sight was out of mind for Edinburgh council, who neglected to provide new housing for the population, resulting in the slums simply moving elsewhere. Plus poor Littlejohn's reforming zeal to improve the habits of the poor as well as their living conditions, led to them calling him 'Wee Hell'. I guess you can't win them all.

By the end of the nineteenth century, Scotland had been sanitised and anglicized by English mores to the extent that the term 'North Britain' was in common usage to describe it. The prime example is the North British Hotel at the east end of Princes Street, which wasn't renamed The Balmoral until the 1980s. A combination of architectural and cultural changes also meant that Edinburgh was losing its unique identity.

The city was desperate to hold onto the veneer of respectability bestowed by its most innovative scholars, scientists, architects and distinguished citizens. Look at the list of notable burials in our nineteenth–century cemeteries at the end of the book, especially Dean. There are a huge number of lesser known advocates, politicians and judges.

Respectable? Yes. Exciting? Not so much.

Edinburgh was being usurped. Glasgow had become the nation's powerhouse and number one city. The highlanders, popularised by George IV's visit and having distinguished themselves in the British Empire's many wars, were regarded with admiration and respect. Now that it was safe to venture up north, its denizens had been labelled 'noble savages' rather than uncouth cattle thieves and everyone mourned the passing of the clans' traditional way of life. The highlanders, not the lowlanders, became custodians of Scotland's character and Bonnie Prince Charlie was transformed from a drunken, ambitious chancer into a national hero.

Edinburgh was suffering a crisis of confidence. It had been a catalyst for many things, but none of it was planned. Now it was in danger of becoming just another generic UK city. Scots was rarely spoken and, though still a seat of learning, we weren't producing thinkers like David Hume, writers like Walter Scott or scientists like James Clerk Maxwell. It became increasingly difficult to make talent count here, for the brightest and best were absorbed by the British Empire, ending up as doctors in India, engineers or missionaries in Africa and the Far East and soldiers all over the world. A rather surprising example of this is the American Civil War Memorial in Old Calton Cemetery

Built in 1893, it commemorates the Scots who fought and died for the Union forces in the American Civil War. It also happens to be the first ever statue of a US president erected outside the USA.

American Civil War memorial, Old Calton.

McGonagall plaque, Greyfriars.

The Forgotten

Despite this mass exodus of go–getters, Edinburgh clung to the notion that it was still a sombre bastion of learning. So it condemmed to obscurity some of the most colourful characters it could have flaunted because they didn't quite fit, or worse, hadn't succeeded in their endeavours.

So let's celebrate a few, starting with the greatest oddball of them all.

William Topaz McGonagall.

There is no denying that Edinburgh has produced some excellent poets but, with the exception of Robert Burns and the Bay City Rollers, it hasn't exactly shaken the world in the Iambic Pentameter department. Yet it houses the grave of the undisputed worst poet in the history of mankind. A close second is James McIntyre, the 'Chaucer of Cheese'. Look him up and be astonished at his ineptitude.

William McGonagall was born (so he claimed) and died in Edinburgh, but worked most of his life as a weaver in Dundee. Thankfully, he didn't compose a poem until the age of fifty–two, for his first foray into the arts was as an actor, a calling he approached with characteristic nuttiness. For instance, playing the lead in *Macbeth*, he felt he was being upstaged by Macduff and refused to die at the end. Then he received a 'divine inspiration' to write, with amazing results. A perfect example of the awfulness he could achieve is encapsulated in this verse from 'The Burns Statue'.

> Fellow citizens, this statue seems most beautiful to the eye,
> Which would cause Kings and Queens for such a one to sigh,
> And make them feel envious while passing by
> In fear of not getting such a beautiful Statue after they die

This 'great poet and tragedian', in his own words, once walked fifty miles to try and convince Queen Victoria to be his patron. When he reached the gates of Balmoral he was told to go away and never come back. He worked in a circus, where the audience were permitted to pelt him with eggs as part of the show. He got an admiring letter from the king of Burma, calling him 'Sir William McGonagall, Knight of the White Elephant' and used the title for the rest of his life, despite the fact that it was a hoax. Though he died penniless in Edinburgh, he remained utterly convinced of his own genius. Perhaps that's why I identify with him so much.

McGonagall has the last laugh, however. Everything he ever wrote is still in publication and he is read avidly today by a core of loyal admirers. His description of The River Tay as 'The Silvery Tay' is always used to describe it and Dundee, not Edinburgh, was savvy enough to claim him as its own.

Then we have Lieutenant John Irvine. He was part of the Franklin Expedition, which

John Irvine memorial, Dean.

set sail in 1845 to find the elusive North West Passage, said to link the Atlantic and Pacific oceans north of Canada. When Franklin's two ships became trapped in pack ice, his men tried to cross the frozen tundra pulling sleds. It was an impossible journey and every man perished. Irvine's grave was finally found on King William Island and he was taken back to Edinburgh and interred in Dean Cemetery.

There is Edinburgh resident Sir Thomas Bouche, who was knighted for designing the first Tay Bridge, the longest structure in the world at the time. In 1879, the bridge collapsed while a train was crossing, killing all seventy–five passengers. The incident was immortalised by none other than William McGonagall in his most famous poem, 'The Tay Bridge Disaster'.

> Beautiful Railway Bridge of the Silv'ry Tay!
> Alas! I am very sorry to say
> That ninety lives have been taken away
> On the last Sabbath day of 1879,
> Which will be remember'd for a very long time.

The collapse was partly due to contractors using shoddy materials and cutting corners but Bouche bore the brunt of public fury and died a broken man. A lucky escape for Edinburgh, perhaps, for he was in the running for the contract to build Edinburgh's Forth Bridge.

How about Sir Hector MacDonald, 'The Fighting Mac'?

Enlisting at seventeen, he ended up a Major General, one of the few men in the British Army who rose through the ranks on merit, rather than achieving the post because he was rich or well connected. He was offered the Victoria Cross for valour in the Afghan Wars and turned it down for a commission in his regiment. Captured during the Boer War, his bravery impressed the enemy so much, they returned his sword when he was released. MacDonald then took part in the expedition to relieve Khartoum and received the Distinguished Service Order for bravery at the battle of Toski. He fought in the Mahdist War and the battles of Abu Hamed, Atbara and Omdurman, where he saved the day by his outstanding military skills. Then back to South Africa for the battles of Paardeberg, Bloemfontein and Pretoria, which gave him a knighthood and made him a household name in Britain. He is even reputed to be the figure on the iconic 'Camp Coffee' labels. This is more than ironic considering what came next.

Why isn't he famous now? Because, despite having a wife and son, he committed suicide after being accused of homosexuality. At least he was spared the embarrassment of having a McGonagall poem written about him.

Perhaps I'm being a bit disingenuous. Who finds lasting fame and who doesn't is, in many ways, pot luck. Here are a few examples.

The Scottish explorer, David Livingstone, is a household name but that's partly due to the fact that he was found living in deepest Africa by the journalist Henry Morton Stanley. Stanley's understated greeting 'Dr Livingstone, I presume,' couldn't fail to go down in history.

Edinburgh University graduate, Mungo Park, who discovered the river Niger, hasn't

Left: Thomas Bouch, Dean.

Right: Hector MacDonald, Dean.

Below: Sir John Murray, Dean.

held onto his fame. Neither has the once celebrated Arctic explorer Robert Goodsir, buried in Dean Cemetery.

Fate can be fickle in other ways. You probably haven't heard of Sir John Murray, even though he is called the 'Father of Oceanography'. He is also interred in Dean Cemetery, which seems the place reputations go to die.

Yet you *do* know him. Murray was the inspiration for Captain Nemo in Jules Verne's *20,000 Leagues under the Sea*.

There is a case to be made that Edinburgh would rather celebrate safe success then glorious failure. At the same time, we became extraordinarily fond of our historical villains. John Porteous. Burke and Hare. Major Thomas Weir executed in 1670 after claiming to be in league with the devil. Maggie Dixon, who survived being hanged in 1724 and went on to open a tavern in the Grassmarket. Jesse King, the 'Stockbridge Baby Farmer', who took in orphans for money then killed them. She was hanged in 1889, buried in Calton Jail and there are no prizes for guessing what's now on top of her grave. Yes. A car park.

All are celebrated in Edinburgh lore, while far more worthy folk have faded into obscurity. A Jekyll and Hyde city indeed.

The Strange Case of Dr Jekyll and Mr Hyde by Robert Louis Stevenson really does capture the strange dichotomies of Edinburgh, embodied in the division between its Old and New Towns, both frequented by the author. To most people, it's just a horror story, but Stevenson's novel is far more than that. I'll go out on a limb and say that it changed the world as much as the Enlightenment.

Unless you've spent your whole life living in the jungles of Borneo, you already know the plot. A respectable Victorian Doctor called Henry Jekyll believes that everyone has a good and bad side, constantly at war with each other. So he invents a potion that separates the two and releases the bad egg known as Mr Hyde, who then goes on a bloody rampage and eventually takes over his maker.

This book about the duality of man was a massive bestseller and read all over the world. It isn't hard to see what it spawned. The werewolf myth, the Incredible Hulk and many superhero movies. Though none can outdo the original which has been filmed more than 125 times.

It also influenced an obscure Austrian scientist called Sigmund Freud, who stopped conducting experiments on eel's testicles, or whatever he was up to, and put forward the theory that each of us is made up of a moral and law abiding superego and a base and savage ID. This kick–started the science of psychology, a globe spanning discipline that its practitioners have been struggling to get right ever since, usually for a minimum of £50 an hour.

Yet a very different movement formed on the back of 'Jekyll and Hyde'. It was inspired by a combination of Darwin's 'Survival of the Fittest' and the Victorians' notion of who the fittest actually were. Scientists and doctors, all from the upper classes, decided they were morally as well as intellectually superior to everyone else. So they figured it was their duty to engineer a way to 'improve' the lower classes. They called the movement 'Social Darwinism' and it caught on big time. Its apex, and logical conclusion, was carried out by the Nazi Party who simply killed everyone they considered inferior. Subsequent attempts at eugenics, genetics or related disciplines have been treated with massive suspicion, though it should be obvious from this short history that human nature could do with a bit of tweaking.

Typically, our major scientific achievement since the Enlightenment is easily the most controversial. The cloning of Dolly the sheep in 1996 took place at the University of Edinburgh's Roslyn Institute.

All of this does Stevenson a disservice. Henry Jekyll isn't a moral man and I think Stevenson was simply exposing the hypocrisy of the Victorians, who would rather improve the character of the underclasses than make their living conditions better.

The Twentieth Century

By the dawn of the twentieth century, The Jekyll and Hyde city was again being pulled in two directions. It could hark back to what it perceived as a glorious past or try to modernise. A great example of looking backwards was the 'Kailyard' school of literature, which is now seen as depicting an overly sentimental view of Scotland's past – conveniently missing out the violence, social problems and grinding poverty.

Its most famous proponent was J. M. Barrie, especially his novel, *A Window in Thrums*. I happen to think that Barrie's work was motivated by sad longing rather than wilful ignorance, and his masterpiece, *Peter Pan*, only reinforces that opinion. At its core, it portrays a boy paying a lonely price for his refusal to move with the times.

On the other hand, there were lauded writers who tried to preserve Scotland's national pride and heritage, like Hugh MacDiarmid and Sorley Mclean, who wrote in Gaelic and was instrumental in preserving its teaching in Scottish schools. My favourite is Lewis Spence (buried in Grange Cemetery), who wrote in classical Scots and helped found what became The Scottish National Party. He was also fascinated by the occult, which is eerily prescient, considering Edinburgh's present reputation as a haunted city.

There is nothing wrong with accepting modernisation, but we often took it way too far. The twentieth century began another round of slum clearance and, in the 1920s and 1930s, families were moved out to barren estates like Niddrie and Prestonfield. This exodus continued until the poorest strata of Edinburgh had been relocated to the periphery, where amenities, civil liberties and transport were virtually non–existent. It didn't solve any intrinsic problems. What it did do was keep them far enough away that tourists would never see the bad parts. That's a process of historical sanitisation that far surpassed anything the Kailyard writers achieved.

Even the finer areas of Edinburgh weren't safe. Historic George Square was practically demolished to make way for Edinburgh University's expansion, while St James' Square was replaced by The St James' Centre, which wouldn't look out of place in industrial Russia. The original buildings of Princes Street were supplanted by a mixture of nasty concrete boxes until, in the 1960s, the historian Moray McLaren called it 'one of the most chaotically tasteless streets in the United Kingdom'.

The relentless march of advancement continued. There were another round of slum clearances in the 1960s and 1970s, shuffling more families to the outskirts where they felt abandoned and resentful. Many still do.

The graveyards suffered just as much from modernisation. Victorian monumental art went out the window and markers became more and more uniform. Advances in cutting and carving machinery, introduced in the 1950s, resulted in the mass production of small stereotypical headstones. The use of native granite virtually ceased and 95 percent of marker materials are now imported. Perhaps the saddest aspect is the regimented rows of identical graves in neat lines, where once they were unique and scattered haphazardly like stone ashes. That's certainly not where I want to end up, so I'm going to be stuffed and kept in a wardrobe.

It's also the reason I've missed out twentieth century cemeteries. Like the city, they have lost much of their personality. In one way, they tell us very little. In another, they show us everything. Unless you are very rich, headstones have become an identikit consumer product, with everyone reduced to buying the same thing. I'm surprised Tesco doesn't sell them.

The old markers aren't immune either. To protect children, the City of Edinburgh Council has taken to pushing over markers if they feel they are unsafe. Since headstones are designed to be upright, the elements then quickly destroy them.

Cemeteries haven't been official playgrounds since the Reformation, so surely it's the parents' responsibility to ensure the safety of their offspring, rather than accept the destruction of historical moments. Seriously. If you want your child to run around somewhere green and safe, take it to a park.

One of the few civic heroes of the twentieth century is Jack Kane, who became the first Labour Lord Provost of Edinburgh in 1972. He was cremated at Warriston

Pushed–over markers, Grange.

cemetery in 1999 and his ashes scattered at the Dean Gallery of Modern Art.

Throughout his career, Kane fought to improve the conditions of deprived communities, using every trick at his disposal. He formed a tenants defence league in the 1930s to oppose evictions and rent hikes. He organised a march of mothers and babies to demand a medical service for the deprived Niddrie area and championed sheltered houses and play groups. He even turned down a knighthood, claiming 'it would separate me from the kind of people I have tried to represent.' When the council finally built the Craigmillar sports and community hall in the 1970s, the locals insisted it be named The 'Jack Kane Centre' as a mark of their esteem. The fact that it looks like a concrete prison from Brasilia can't detract from that sentiment.

Edinburgh Today

Little did the city realise that a lifeline to its civic pride was gaining in strength. Jack Kane was an enthusiastic supporter of the Edinburgh Festival and Fringe, founded in 1947, which has grown to become the world's largest celebration of the Arts. In 2012, it boasted over 2,500 shows from forty–seven countries in 279 venues, and it continues to expand. It launched the careers of Alan Bennet, Peter Cook, Dudley Moore, Jonathan Miller, Derek Jacobi, Monty Python, Tom Stoppard, Craig Ferguson and countless others, none of whom actually come from Edinburgh. (Ferguson is from Glasgow).

Jack Kane Centre, Craigmillar.

The city had found a new source of international acclaim and it capitalised on it. Science, architecture and philosophy were old hat. Edinburgh was now a bastion of the arts.

In a way, this was always on the cards. Look at the list of dead people at the end of the book. How many artists? Sculptors? Writers? Plenty. Now they were reclaimed and revered as a vital part of the city's history and a pointer to its future.

Paintings by Edinburgh's 'Scottish Colourists' Samuel Peploe, John Duncan Fergusson and Francis Cadell sell for huge amounts, despite the fact that Cadell died in poverty as recently as 1937 (he and Peploe are buried in Dean). Related artists like Samuel Bough (buried in Dean, 1878) and the Scottish pioneer of Impressionism, William McTaggart (buried in Newington, 1910) are also highly regarded. The Victorian sculptor John Steell, whose statues are dotted all over Edinburgh, is regarded as a master of the equine form, even though he's modestly buried in an unmarked grave in New Calton.

As befits the Jekyll and Hyde City, our prime exponent of art is now Jack Vettriano. Though his work is lambasted by critics as 'brainless erotica', his 1992 painting 'The Singing Butler' is the best selling art print in Britain.

Writers like Arthur Conan Doyle, Muriel Spark, Nigel Trantor, Sorley MacLean, R. M. Ballantyne, J. M. Barrie, George Mackay Brown, Dorothy Dunnett, Kenneth Grahame, Norman McCaig, Compton Mckenzie, and Sorley McLean, had always been popular but now they proved the vanguard of a writing renaissance. In 2004, Edinburgh became the first UNESCO City of Literature and a new generation of authors, who were born or lived in the city, gained acclaim and fame. Iain Banks, Irvine Welsh, Ian Rankin, Lin Anderson, Aileen Paterson, J. K. Rowling, Ian McCall Smith, Iain Banks and, of course, me. Though I've managed to avoid the famous part.

However, their relationship with the capital isn't the same as writers like Walter Scott or Robert Louis Stevenson, who regarded Edinburgh so fondly. Ian Rankin fills it with criminals, as does Irvine Welsh. Quentin Jardine's *Skinner* series and Christopher Brookmyer's *Quite Ugly One Morning* positively revel in the malicious side of the capital. Jonathan Aycliffe's Faustian chiller *The Matrix* and James Robertson's *The Fanatic* portray the city as a malevolent entity lurking under a thin layer of sophistication. Edinburghers snigger that part of the movie *Trainspotting* had to be filmed in Glasgow because our housing estates looked too nice, conveniently forgetting that Irvine Welsh was chronicling the very real heroin culture that thrived here. They are also proud of Spark's *The Prime of Miss Jean Brodie*, despite the fact that Brodie's Edinburgh teacher is a wilfully ignorant, hypocrite. We're immensely chuffed by Eleanor Atkinson's wildly inaccurate children's book, *Greyfriars Bobby*, even though she was an American who never set foot in the city. Even J. K. Rowling's adult novels take place in England, where Harry Potter also lived, before being sent to the mythical Hogwarts.

However, the graveyards still exert their lure.

Rowling wrote part of the first *Harry Potter* book in the Elephant House Café overlooking Greyfriars. What can you see from the window? On the other side of the graveyard is Herriot's School, which looks like a castle and used to be an orphanage. Hogwarts? Right below the window is narrow Candlemaker Row which, at the time,

Herriot's School, looking over Greyfriars.

contained a witchcraft shop. Diagon Alley? I have a friend who swears he found a grave in Greyfriars bearing the name Tom Riddle (Lord Voldemort in the Harry Potter books) but I haven't been able to find it. Perhaps you'll have better luck.

This brings us to the present day. What exactly is Edinburgh now?

There's a great phrase to describe the city, 'All fur coat and no knickers', which you usually hear from Glaswegians. It basically means superficially elegant but actually quite common. I think 'common' is the wrong word. 'Venal' is more accurate. Or perhaps 'cunning'.

Edinburgh has enjoyed too much fame to slide into obscurity, so it is determined to have its cake and eat it too. That's our real achievement as we move into the twenty–first century. That's the Jekyll and Hyde city at its finest.

We have become all the things visitors want us to be. Walk up the Royal Mile and you are bombarded by tartan tat, mini–kilts and hats with red hair sticking out, much loved by stag parties. That's really the highlands, the area we spent so long despising and fighting.

Despite our Calvinist heritage, we have become known as a party city and welcome these stag and hen groups, even though residents hate them. Our Hogmanay celebrations are some of the largest in the world, attracting thousands of visitors who don't have a clue about the customs associated with the celebration. So we put on fireworks and concerts instead of first footing, shaking hands and kissing complete strangers, which is much more fun. Still, it brings in the money.

We have a carefully nurtured reputation as one of the most haunted cities on earth, packed with ghost tours. That status, and none of the tours, existed before the 1980s.

Presbyterian St Giles' Kirk is always called a St Giles' Cathedral, though it only held the once despised title for a few years under Charles I. It must sound more impressive to visitors. The home of the Reformation and the Covenanters even has Beltane and Halloween Pagan festivals, including a procession which starts right next to John Knox's plaque.

Because of the Festival and Fringe, we are celebrated as a theatre city. Yet the Fringe is mostly comedy acts, only lasts for four weeks and Edinburgh isn't exactly overflowing with theatres – especially compared to Glasgow.

The Jekyll and Hyde City. Is it something we should applaud or lament? Is it something we planned, or has it been foisted on the population? I honestly don't know, so I'll leave the last words to Jack Kane, for they'll always be true.

'Too many people are prepared to let things happen to them instead of taking part in deciding what is happening'.

Eventually we all die. There are graveyards and markers waiting for everyone. When you go to Greyfriars or Colinton or Canongate, look at these monuments in a different light. Think about who these people were and what they achieved. Yes, they added to Edinburgh's rich tapestry, but that wasn't in the forefront of their minds. They simply had thoughts and ambitions and feelings, like you and I.

They were real. They existed. Standing over their last resting place is the closest you will ever get to these individuals. So pay your respects and be encouraged.

You, after all, are still alive.

There's time to make your own mark on history.

List of Burials

Holyrood Abbey

Notable Burials

Sibilla de Stratun (Stone slab dated c.1300).

King David II (1324–71).

King James II (1430–1460).

Mary of Guelders (1434–1463). Wife of James II and originally buried in Trinity College Kirk.

Arthur, Duke of Rothesay (1509–1510). Second son of James IV.

Madeleine of Valois (1520–1537). Wife of James V.

James, Duke of Rothesay (1540–1541). Eldest son of James V.

Arthur, Duke of Albany (1541–1541). Second son of James V, who died in infancy.

King James V (1512–1542).

Henry Stuart, Lord Darnley (1545–1567).

George Douglas, Bishop of Moray (d.1589).

Margaret Stewart (1598–1600). Second daughter of James VI.

Robert Douglas, Viscount Belhaven (d.1639).

Alexander Mylne (d. 1643). Master mason.

Bishop George Wishart (1599–1671).

George, Earl of Sutherland (1633–1703).

Dunbar Douglas (1722–1799). 4th Earl of Selkirk.

St Cuthbert's

Notable Burials

Henrie Nisbet of Dean (d.1592). Buried underneath the church.

John Napier (1550–1617). Mathematician and inventor of logarithms.

Rev. David Williamson (1636–1706). Covenanter and Moderator of the General Assembly of the Church of Scotland.

Charles Darwin (1758–1778). Uncle of Charles Darwin.

Alexander Murray, Lord Henderland (1736–1795). Judge and politician.

John Cunningham, 15th Earl of Glencairn (1749–1796). Cavalryman and priest.

Thomson Bonar (1739–1814). Co-founder of

Encyclopaedia Britannica.

Girolamo Stabilini (1761–1815). Violinist.

Adam Rolland (1734–1819). Clerk of Session and early fashionista.

Thomas Morison (1761–1820). Builder of large sections of the New Town.

Dr Henry Dewar (1771–1823). Minister, physician and writer.

Sir Henry Raeburn (1756–1823). Famous artist.

Reverend Robert Smith (1780–1829). Composer of hymns, including 'How Beautiful Upon the Mountains'.

Robert Jamieson (1784–1834). Lawyer. Monument by the famous sculptor John Steell.

George Watson RSA artist (1767–1837) and son William Smellie Watson (1796–1874). Artists.

John Jamieson (1759–1838). Author of *Etymological Dictionary of the Scottish Language.*

Anne Grant (1755–1838). Author of *Letters from the Mountains.*

Alexander Nasmyth (1758–1840). Artist. Buried with his sons Patrick and James.

John Abercrombie (1780–1844). Physician.

James Stevenson RSA (1780–1844). Artist.

George Meikle Kemp (1795–1844). Architect and designer of the Scott Monument.

Andrew Combe (1797–1847). Phrenologist.

John Stark FRS (1779–1849). Author and printer.

Susan Ferrier (1782–1854). Novelist.

Thomas De Quincey (1785–1859). Author of *Confessions of an English Opium-Eater* and an influence on Edgar Allan Poe.

James Pillans (1778–1864). Educator.

James Frederick Ferrier (1808–1864). The first epistemologist.

George Aikman (1788–1865). Engraver.

James Craufurd, Lord Ardmillan (1804–1876). Law lord.

David Rhind (1808–1883). Architect.

Duncan McLaren (1800–1886). Lord Provost and son, Walter (1853–1912). MP.

Robert Traill Omond (1858–1914). Physicist and geologist.

Walter Biggar Blaikie (1847–1928). Engineer, historian and astronomer.

The heart of Scottish born, Canadian sculptor Robert Tait McKenzie (1867–1938), despite his request that it be buried in Calton Cemetery.

Greyfriars

Notable Burials

James Douglas, 4th Earl of Morton (d. 1581). Regent of Scotland.

George Buchanan (d. 1582). Historian and Reformer.

Alexander Henderson (d. 1646). Churchman and statesman.

John Mylne (1611–1667). Architect.

Archibald Campbell, 9th Earl of Argyll (1629–1685). Nobleman.

Sir George Mackenzie (1636–1691). Lord Advocate. His tomb is by James Smith and is modelled on the Tempietto di San Pietro.

Mary Erskine (1629–1708). Founder of The Mary Erskine School.

Archibald Pitcairne (1652–1713). Physician.

William Carstares (1649–1715). Churchman and statesman.

George Watson (1654–1723). Accountant and founder of George Watson's College.

Colonel Francis Charteris (1675–1732). Rake.

Captain John Porteous (*c.* 1695–1736). Soldier.

Colin MacLaurin (1698–1746). Mathematician.

Duncan Forbes, Lord Culloden (1685–1747). Politician and judge.

William Adam (1689–1748) and his son John Adam (1721–1796). Architects.

Thomas Ruddiman (1674–1757). Classical scholar and grammarian.

Allan Ramsay (1686–1758). Poet.

James Stirling (1692–1770). Mathematician.

Andrew Crosbie (1736–1785). Founding Fellow of the Royal Society of Edinburgh.

William Robertson D.D. (1721–1793). Historian.

James Craig (1739–1795). Architect and designer of Edinburgh's New Town.

William Smellie (encyclopedist) (1740–1795). Editor

of the *Encyclopaedia Britannica*.

James Hutton (1726–1797). Geologist.

Joseph Black (1728–1799). Physician.

William Coulter (1808–1810). Lord Provost.

Duncan Ban MacIntyre (1724–1812). Famed Gaelic poet.

Patrick Miller of Dalswinton (1731–1815). Steamship innovator.

William Creech (1745–1815). Bookseller and Lord Provost.

John Kay. (d.1824). Caricaturist.

Henry Mackenzie (1745–1831). Author of *The Man of Feeling*.

Thomas M'Crie (1772–1835) and his son Thomas M'Crie (1797–1875). Ministers.

Major General William Farquhar (c.1770–1839). First president of Singapore.

William Wallace (1768–1843). Mathematician.

Prof George Dunbar (1777–1851). Classical scholar.

John Grey (d1858). Bobby's master.

William McGonagall (1825–1902). The world's worst poet.

Greyfriars Bobby. A dog.

Monuments

The Martyr's Monument, which commemorates executed Covenanters.

Monument to Duncan Ban MacIntyre. Renovated in 2005.

Monument of John Byres of Coates by the master mason William Wallace.

Greyfriars Bobby headstone and statue. (The statue is just outside).

Canongate

Notable Burials

John Frederick Lampe (1703–1751). Composer.

Bishop Robert Keith (1681–1757). Author.

Professor Charles Alston (1683–1760). Lecturer and co-founder of the Edinburgh School of Medicine.

George Drummond (1688–1766). Six times Lord Provost and founder of Edinburgh's New Town.

John Gregory MD (1724–1773) and son James Gregory MD (1753–1821). Scientists.

Robert Fergusson (1750–1774). Poet.

Daniel Dow (1732–1783). Fiddler and composer.

Alexander Runciman (1736–1785) and brother John Runciman (1744–1766). Painters.

John Mackenzie, Lord MacLeod (1727–1789). Jacobite and Swedish Count.

Adam Smith (1723–1790). Economist and author of *The Wealth of Nations*.

Rev. Thomas Hardy (1748–1798). Professor of Church History and Divinity.

Benjamin Bell (1749–1806) and his son Joseph Bell, (1787–1848). Surgeons.

James Clark (1732–1808). Founded of the James Clark Vetinary School.

Mary Brunton (1778–1818). Novelist.

Luke Fraser (1736–1821). Teacher of Sir Walter Scott.

Dugald Stewart (1753–1828). Professor of Moral Philosophy at Edinburgh University and author of *Philosophy of the Human Mind*.

Hugh William Williams (1773–1829). Watercolourist and landscape artist.

John Ballantyne (1774–1821), and his brother James Ballantyne (1772–1833). Publishers of Sir Walter Scott's *Waverley* novels.

Sir William Fettes (1750–1836). Lord Provost and founder of Fettes College.

George Chalmers (1773–1836). Plumber and founder of Chalmers Hospital.

Mrs Agnes Maclehose (1759–1841). Robert Burns' 'Clarinda'.

Euphemia Amelia Murray (1768–1845). Burns' 'Flower of Strathmore'.

Sir John Watson Gordon R.A. (1788–1864). Artist.

Horatius Bonar (1808–1889). Preacher and hymn-writer.

William McEwan (1827–1913). Founder of McEwan's Brewery.

Robert Hurd (1905–1963). Architect.

Reputed Burials and Unmarked Graves

David Rizzio (1533–1566). Musician.

James Ramsay (1624–1696). Bishop of Ross.

Nicolo Pasquali (died 1757). Musician.

William Wilson (1709–1815). Jacobite and Colonel of the Black Watch.

David Douglas, Lord Reston (1769–1819). Judge.

John Schetky (1740–1824). Composer and co-founder of the Boar Club.

Alexander Campbell (1764–1824). Jacobite and publisher of Scots songs.

Ebeneezer Scroggie (1792–1836). Inspiration for Ebeneezer Scrooge.

Monuments

Memorial Cross for the soldiers of Edinburgh Castle. The dead lie in the wide open green area around the cross.

The Coachman's Stone. A memorial dedicated to the drivers operating the Edinburgh to London route. Several of the Company are interred next to it.

Old Calton

Notable Burials

William Woods (d 18th century). Actor and friend of Roberts Fergusson and Burns.

Members of the Haig family, who invented blended whisky and turned it into an industry.

David Hume (1711–1776). Philosopher and author of *Treatise of Human Nature*. Mausoleum by Robert Adams.

David Allan (1744–1796). Painter known as 'The Scottish Hogarth'.

Peter Williamson (1730–1799). White slave.

Daniel Stewart (1741–1814). Founder of Daniel Stewart's Hospital, which eventually became Stewart's Melville College.

Robert Burn (d. 1815). Architect, including Nelson's monument next door.

Professor John Playfair (1748–1819). Mathematician and scientist.

Archibald Constable (1774–1827). Publisher who founded the *Edinburgh Review* and also published Scott's work.

Julius Von Yelin (d. 1826). German Knight and scholar.

William Blackwood (1776–1834). Publisher, magistrate, founder of the *Edinburgh Encyclopaedia*, *Blackwood's Edinburgh Magazine* and Sir Walter Scott's publisher.

Charles Mackay (d. 1857). Comedian and actor. Allegedly the person referred to by the phrase 'the real McCoy'.

Richard Dickson (1792–1857). Architect.

Thomas Hamilton RSA (1784–1858). Architect of the Royal High School, the Bank of Scotland on the Mound, Physicians Hall, the Dean Gallery and the Political Martyrs Monument, just a few yards north of his grave.

Prof. George Wilson MD (1818–1859). Author and biographer.

James Leishman McDougal (1840–1869). Soldier. Awarded the Victoria Cross for his infiltration of Taku Forts during the Second China War.

Dr Robert Candlish (1806–1873). Clergyman and leader of the Disruption of the Scottish Church.

Sir John Steell (1804–1891). Acclaimed Scottish sculptor. Unmarked grave.

James Lumsden (1836–1899). Singer.

Monuments

Political Martyrs' Monument, The American Civil War Memorial and a number of eighteenth-century monuments to local tradesmen.

New Calton

Notable Burials

Archibald Elliot (1760–1823). Architect.

Rear Admiral James Bisset (d. 1824). Sea Captain of the Cunard Line.

William Knox (1789–1825). Abraham Lincoln's favourite poet.

Vice Admiral Alexander Frazer (1747–1829).

Andrew Skene (1784–1835). Solicitor General for Scotland.

Robert Stevenson (1772–1850). Civil engineer.

General Roger Hale Sheaffe (1763–1851). Loyalist in the Anglo–American War.

William MacGillivray (1796–1852). Naturalist.

Admiral John Graham (1791–1854).

Robert Pitcairn (1793–1855). Antiquary.

Alan Stevenson (1807–1865). Lighthouse engineer.

James Ivory, Lord Ivory (1792–1866). Judge.

Alexander Bryson (1816–1866). Scientist.

Dr James Begbie (1798–1869). President of Royal College of Physicians.

Vice Admiral Thomas Frazer (1796–1870).

David Bryce (1803–1876). Architect.

Robert Christison (1797–1882). Toxicologist.

Dr John Brown (1810–1882). Author.

Thomas Stevenson (1818–1887). Lighthouse engineer.

John Inglis, Lord Glencorse (1810–1891). Politician and judge.

Monuments

The Commonwealth War Grave. Dedicated to five merchant navy seamen whose bodies were recovered from the sea following an attack on the *MV Atheltemplar* during the Second World War.

St John's

Notable Burials

Anne Rutherford (1739–1819). Mother of Sir Walter Scott.

James Donaldson (1751–1830). Founder of Donaldson's School for the Deaf.

Thomas Balfour (1810–1838). MP for Orkney and Shetland.

Bishop James Walker (1770–1841). Anglican minister.

Lesley Baillie (1768–1843). Yet another subject of a Robert Burns' poem, this time 'Bonnie Lesley'.

General Sir Archibald Campbell (1769–1843). Commanded British forces in the first Anglo Burmese–War and became Lieutenant Governor of New Brunswick.

William John Thomson (1771–1845). Artist.

James Walker (1781–1862). Civil engineer.

Margaret Outram (1778–1863) wife of Benjamin Outram (1768–1805), pioneer of canal building and the mother of Sir James Outram (1803–1863) who modestly turned down a Victoria Cross for his valour in the Indian Rebellion of 1857.

General Anthony MacRae (1812–1868).

Dean Edward Bannerman Ramsay (1793–1872). His grave is separate from the monument to him, which was designed by Robert Rowand Anderson.

Prof Thomas Laycock (1812–1874). Physiologist.

Daniel Fox Sandford (1831–1906). Bishop of Tasmania.

George Young, Lord Young (1819–1907). Liberal MP and judge.

William Campbell, Lord Skerrington (1855–1927). Senator of the College of Justice.

Bishop Harry Reid (d. 1943). Episcopal Bishop of Edinburgh.

Memorials

General Sir John Campbell (1807–1855). 2nd Baronet of New Brunswick, Canada and son of General Sir Archibald Campbell, who is buried nearby.

Dean Edward Bannerman Ramsay (1793–1872). Clergyman and author of *Reminiscences of Scottish Life and Character*.

Warriston

Notable Burials

Alexander Ramsay (1777–1847). Architect.

Feliks Janiewicz (1762–1848). Exiled Polish composer and violinist.

William Nicol (1770–1851). Physicist and geologist.

Patrick Neill (1776–1851). Naturalist.

Professor Robert Jameson (1774–1854). Naturalist and mineralogist.

Count Walerian Krasiński (1795–1855). Polish politician and historian.

John Siveright (1779–1856). Fur trader and Chief Factor of the Hudson Bay Company.

Thomas Jameson Torrie (d. 1858). Advocate, geologist, mineralogist and botanist.

James Jardine (1776–1858). Civil engineer.

Professor David Low (1786–1859). Agriculturalist.

Captain Francis Stupart (1781–1860). Cavalry Officer at the Battle of Waterloo.

Horatio McCulloch (1806–1867). Artist. Monument by John Rhind.

Alexander Smith (1829–1867). Poet.

Robert Scott Lauder (1803–1869). Artist. Monument by John Hutchison.

James Eckford Lauder (1811–1869). Artist.

Sir James Young Simpson (1811–1870). Pioneer of anaesthetics.

Bishop Charles Hughes Terrot (1790–1872). Episcopal minister and mathematician.

Stewart McGlashan (1807–1873). Sculptor.

Adam Black (1784–1874). Publisher, Lord Provost and MP for Edinburgh.

Duncan McNeill, 1st Baron Colonsay and Oronsay (1793–1874). Advocate, Tory politician, Lord Justice General and Lord President of the Court of Session.

Sir George Harvey (1805–1876). Artist.

Sir David Deas (1807–1876). Naval physician.

Charles Neaves, Lord Neaves (1800–1876). Judge.

Samuel Halkett (1814–1877). Librarian and author.

Hugh Morton (1812–1878). Civil engineer.

Philip Kelland (1808–1879). Mathematician.

John Menzies (1808–1879). Founder of the national newsagent chain.

Robert MacFarlane, Lord Ormidale (1802–1880). Judge.

William Robertson (1818–1882). Physician and statistician.

Robert Gavin (1827–1883). Artist.

Sir George Harrison (1812–1885). Lord Provost.

James Pringle (1822–1886). Businessman and Provost of Leith.

John Dick Peddie (1824–1891). Architect.

Alexander Keiller (1811–1892). Physician, obstetrician and gynaecologist.

John Rhind (1828–1892). Sculptor.

Alexander Nicolson (1827–1893). Scholar and mountaineer.

Alexander Henry (1818–1894). Gunsmith and Town Councillor.

Sir John Struthers (1823–1899). Surgeon and anatomist.

John Smart (1838–1899). Landscape artist.

Sir Thomas Clark (1823–1900). Lord Provost.

William Williams (1832–1900). Veterinary surgeon and founder of Dick Vet College.

Thomas Menzies (1847–1901). Shipbuilder.

George Aikman (1830–1905). Artist and engraver.

Alexander Peddie (1810–1907). Physician and author.

John Cumming (1824–1908). Artist.

John Smith (1825–1910). Surgeon and founder of the Edinburgh Dental Hospital.

Sir John Batty Tuke (1835–1913). MP and psychiatrist, with a fitting middle name.

Hippolyte Blanc (1844–1917). Architect.

Robert Rowand Anderson (1834–1921). Architect.

Andrew Grant (1830–1924). Liberal MP.

Sir William Peck (1862–1925). Astronomer.

Harold Raeburn (1865–1926). Mountaineer.

James Cadenhead (1858–1927). Artist.

William Graham Boss (1883–1927). Stained glass designer.

Robert William Dundas MC (1881–1928). Military hero who was awarded the Legion of Honour before becoming a solicitor.

Robert Gibb (1845–1932). Artist who painted 'The Thin Red Line'.

Malcolm Smith (1856–1935). MP and Provost of Leith.

Sir John James Burnet (1857–1938). Architect.

Sir Louis Stewart Gumley (d. 1941). Lord Provost.

Lorne MacLaine Campbell (1902–1991). Awarded the Victoria Cross for his actions during the Second World War.

Cremated

The crematorium is on a separate site, east of the main cemetery.

Sir Robert Lorimer (1868–1929). Architect who designed the war memorials in Edinburgh Castle and the Thistle Chapel in St Giles' Cathedral.

Alfred Adler (1870–1937). Austrian psychotherapist.

Ebenezer James MacRae (1881–1951). City Architect.

Captain Charles George Bonner (1884–1951). Awarded the Royal Navy Victoria Cross during the First World War when his ship was attacked by an enemy submarine.

Sir Frank Mears (1880–1953). Architect and town planner.

Andrew Gilzean (1877–1957). MP.

Sapper Adam Archibald (1879–1957). Awarded the Victoria Cross for valour at the battle of Sambre, during the First World War.

Captain Henry Peel Ritchie (1879–1958). Awarded the Royal Navy Victoria Cross during the First World War, for refusing to leave the helm of his beleaguered ship, despite being badly injured, until he had steered it to safety.

Drum–Major Walter Potter Ritchie (1892–1965). Awarded the Victoria Cross at the Battle of the Somme, for rallying his men in full view of enemy machine gun fire.

Lieutenant David Lowe MacIntyre (1895–1967). Awarded the Victoria Cross during the First World War, for single-handedly charging and capturing a machine gun post.

Sir Charles Laing Warr (1892–1969). Minister of St Giles.

Brigadier Arthur Edward Cumming (1896–1971). Awarded the Victoria Cross during the Second World War, defending an airfield during the Battle of Malaya.

Anthony Chenevix–Trench (1919–1979). Headmaster of Eton and Fettes Colleges.

Tom Hart (1922–1982). Chairman of Hibernian Football Club.

Don Revie (1927–1989), English footballer and manager.

Jack Kane (1911–1999). Lord Provost.

War Graves

Warriston Cemetery contains 100 graves of Commonwealth service personnel, 72 from World War One and 27 from World War Two. The cemetery also contains a Commonwealth War Graves Commission memorial, in the form of panels, listing 142 cremated World War Two Commonwealth service personnel.

Dean

Noteable Burials
Original Cemetery

Sir Ludovic Grant (1862–1836). 11th Baronet of Dalvey.

Rev Angus Makellar (d. 1859). Moderator of the Church of Scotland.

Arthur Perigal RSA (1784–1847). Artist.

Lt John Irving (1822–1848 or 49). Explorer.

John Peter Grant (1774–1848). MP.

Sir William Allan RSA (1782–1850). Artist.

Lord Francis Jeffrey (1773–1850). Judge and Literary critic.

Sir James Wellwood Moncrieff, 9th Baronet, Lord Moncrieff (1776–1851). Judge.

Thomas Thomson (1768–1852). Advocate.

Lord Andrew Rutherfurd (1791–1852). Advocate, judge and politician.

Lord Henry Cockburn (1779–1854). Judge.

Prof John Wilson (1785–1854). Author who wrote under the name Christopher North.

Prof Edward Forbes (1815–1854). Naturalist.

William Henry Playfair (1790–1857). Famed architect, who also designed some of the Dean monuments.

Prof John Fleming (1785–1857). Naturalist.

George Combe (1788–1858). Lawyer and phrenologist.

Alexander Black (1797–1858). Architect.

William John Gairdner, CB, (1789–1861). Major General of the Bengal Infantry.

James Haliburton (1788–1862). Egyptologist.

William Ambrose Morehead (1805–1863). Governor of Madras.

Prof William Edmondstoune Aytoun (1813–1865). Poet.

Prof Henry Darwin Rogers (1808–1866). Geologist.

Sir Archibald Alison (d.1867). Advocate and historian.

John Goodsir (1814–1867). Anatomist.

John Stevens (1798–1868). Artist.

Prof James David Forbes (1809–1868). Inventor of the seismometer.

John Ritchie (1778–1870) and John Ritchie Findlay

(1824–1898). Newspaper tycoons.

David Octavius Hill (1802–1870). Artist and photography pioneer. The monument is by his sculptor wife Amelia Robertson Hill (1820–1904), who is buried with him.

Edward Maitland, Lord Barcaple (1803–1870). Advocate and judge.

Alexander Russel (1814–1870). Editor of *The Scotsman*.

Thomas Bonnar (d.1873) (d.1896). Father and Son. Artists and designers.

Robert William Thomson (1822–1873). Engineer and inventor of the pneumatic tyre.

John Lisle Hall MacFarlane (1851–1874). Scotland rugby international.

Andrew Inglis (d. 1875). Fellow of the Royal College of Surgeons of Edinburgh, and Professor of Midwifery at Aberdeen University.

Donald Mackenzie (1818–1875). Judge.

John McEwan (1832–1875). Brewer of the famous McEwan's Lager.

Robert Matheson (1808–1877). Architect.

George Paul Chalmers (1838–1878). Artist.

David Cousin (1809–1878) Architect. His name is recorded on the family stone though he's buried in the USA.

Samuel Bough RSA (1822–1878). Artist.

Edward Gordon, Baron Gordon of Drumearn (1814–1879). Judge and politician.

John Blackwood (1818–1879). Creator and editor of *Blackwood's Magazine*.

Memorial to George Brown (1819–1880). Canadian politician.

Sir Thomas Bouch (1822–1880). Railway engineer.

Vice Admiral Charles Fellowes (1823–1880).

John Hill Burton, Isabella Burton and family (d. 1881) Historian and Author. Burton isn't actually in the plot, as he was divorced when he died.

William Brodie (1815–1881). Sculptor, who designed some of the monuments in Dean Cemetery.

Prof William Rutherford Sanders (1828–1881). Pathologist.

John Miller (1805–1883). Railway and dock engineer.

John Lessels (1808–1883). Architect.

Sir Henry Wellwood–Moncreiff, 10th Baronet (1809–1883). Minister.

Dr Alexander Wood (1817–1884). Inventor of the hypodermic syringe.

Cunninghame Borthwick (1813–1885). 19th Lord Borthwick.

James Young Gibson (1826–1886). Author and translator.

John Waddell (1828–1888). Railway engineer.

Robert Chambers (1832–1888). Publisher of *Chambers Encyclopedia*.

Thomas Stuart Burnett (1853–1888). Sculptor.

Sir James Falshaw (1810–1889). Lord Provost.

Lord Patrick Fraser (1817–1889). Dean of the Faculty of Advocates.

James Nasmyth (1808–1890). Inventor of the steam hammer.

John Crabbie (1806–1891). Founder of 'Crabbie's Green Ginger Wine'.

Sir William Fettes Douglas (1822–1891). Artist.

James Hamilton, (1822–1893). 9th Baron of Belhaven and Stenton.

Thomas Drybrough (1820–1894). Brewer.

Charles Kinnear (1830–1894). Architect.

John Stuart Blackie (1809–1895). Scholar.

Sir William Fraser (1816–1898). Historian.

William Hamilton Beattie (1842–1898). Architect who designed Jenner's department store and the Balmoral Hotel.

Robert Anstruther Goodsir (1823–1899). Doctor and Arctic explorer.

Robert Cox (1845–1899). MP.

William Watson, Baron Watson (1827–1899). Law lord.

Robert Carfrae (1820–1900). Antiquarian.

Prof Sir Thomas Grainger Stewart (1837–1900). Physician.

Robert Younger (1820–1901). Brewer and creator of 'Younger's Tartan Special'.

Joseph Noel Paton (1821–1901). Artist.

Duncan Cameron, (1825–1901). Newspaper owner and inventor of the 'Waverley' pen. Buried with his daughter, Mary Cameron (1865–1921). Painter.

Major General Sir Hector MacDonald, (d. 1903).

War hero.

Sir James Steel (1830–1904). Lord Provost.

Isabella Bird (1831–1904). Traveller, writer, photographer and first female Fellow of the Royal Geographical Society.

Sir William Muir (1819–1905). Scottish Orientalist.

Alexander William Black (1859–1906). MP.

John James Stevenson (1831–1908). Architect.

Professor Daniel John Cunningham (1850–1909) and his son General Sir Alan Cunningham (1887–1983).

Lord Charles Pearson (1843–1910). Law lord.

Samuel Butcher (1850–1910). President of the British Academy and Liberal Unionist MP.

Joseph Bell (1837–1911). Personal surgeon of Queen Victoria.

Alexander Taylor Innes (1833–1912). Lawyer and historian.

Sir James Gibson (1849–1912). Lord Provost and MP for Edinburgh.

Rev Cameron Lees (1835–1913). Minister and author.

Sir John Murray (d. 1914). Father of Oceanography.

Dr Henry Littlejohn (1826–1914). Public health promoter and forensic science pioneer. He is buried with his son, Henry Harvey Littlejohn (1862–1927), also a forensic scientist and Edinburgh's first Police Surgeon.

Thomas Clouston (1840–1915). Psychiatrist.

Edward and James Key (b. 1889) (d. 1915). Jute barons and philanthropists.

Sir Stair Agnew (1831–1916). Baronet.

Rev Alexander Robertson MacEwen (1851–1916). Minister and author.

Sir William Turner (1852–1916). Anatomist.

Henry Martyn Clark (1887–1916). Missionary.

Baron Kinnear (1833–1917). Advocate and judge.

Elsie Inglis (1864–1917). Pioneering doctor and war hero.

Benjamin Hall Blyth (1849–1917). Civil engineer.

Sir Mitchell Mitchell–Thomson (1816–1918). Lord Provost.

Sir James Russell (1846–1918). Lord Provost.

Sir Andrew Henderson Leith Fraser (1848–1919). Lieutenant Governor of Bengal.

Baron Charles John Guthrie (1849–1920). Law lord.

John More Dick Peddie (1853–1921). Architect.

Alexander Ignatius Roche (1861–1921). Artist.

Alexander Crum Brown (1838–1922). Chemist.

Edward Arthur Walton (1860–1922). Artist.

Thomas Drummond Wanliss (1830–1923). Journalist, author and politician in Australia, who died after returning to Scotland.

John Abercromby, 5th Baron Abercromby (1841–1924). Soldier and archaeologist.

John Sinclair, 1st Baron Pentland (1860–1925). Soldier and politician.

Sir Thomas Hutchison (1866–1925). Lord Provost.

Robert Alexander RSA (1840–1927). Artist.

Henry Snell Gamley (1865–1928). Artist.

Prof George Gregory Smith (1865–1932). Literary critic.

Walter Oliphant (1867–1933). Publisher.

Samuel Peploe (1871–1935). Artist.

Sir Frederick Thomson, 1st Baronet (1875–1935) and his son Sir Douglas Thomson, 2nd Baronet (1905–1972). Politicians.

Robert Gemmell Hutchison (1855–1936). Artist.

Henry Wright Kerr RSA RSW (1857–1936). Artist.

Lord David Fleming (1877–1944). Military hero.

Victor Noel–Paton, Baron Ferrier (1900–1992). Soldier and businessman.

John Bellany (1942–2013). Artist.

Twentieth–Century Extension

Arthur Dewar, Lord Dewar (1860–1917). Judge and Liberal politician.

Charles Scott Dickson, Lord Dickson (1850–1922). Judge and politician.

Joseph Shield Nicholson (1850–1927). Economist.

Andrew Constable, Lord Constable (1865–1928). Judge and Tory politician.

William Skeoch Cumming (1864–1929). Artist.

Sir John Ritchie Findlay (1866–1930). Newspaper magnate.

Alexander Munro MacRobert (1873–1930). MP and Lord Advocate.

Edward Theodore Salvesen, Lord Salvesen (1857–1942). Politician and judge.

Douglas Strachan HRSA (1875–1950). Stained glass

window designer.

Stewart Kaye (1886–1952). Architect.

Herbert John Clifford Grierson (1866–1960). Editor and literary critic.

Sydney Goodsir Smith (1915–1975). Poet and artist.

Sir Alexander MacPherson Fletcher (1929–1989). MP.

Herrick Bunney CVO (1915–1997). Organist at St Giles' Cathedral.

Sir David William Scott–Barrett (1922–2003). Soldier and commander of the British section in Berlin.

Monuments

Monument to John George Bartholomew, a map–maker buried in Portugal, sculpted by Pilkington Jackson.

Monument to the 79th Cameron Highlanders marking their role in the Crimean War at Alma and Sevastapol and their part in the Indian Mutiny at Lucknow.

Monument to the Edinburgh–born Confederate Colonel Robert A. Smith, who died in 1862 at Munfordsville, Kentucky, during the American Civil War.

Monument to historian John Hill Burton, buried at Dalmeny.

Monument to vocalist, John Wilson, buried in Quebec.

The war graves of Commonwealth Service Personnell.

Rosebank

Notable Burials

James Slight (1785–1854). Lighthouse engineer.

James Bertram (1816–1861). Engineer.

Henry Stephens (1794–1874). Author of *The Book of the Farm*.

Andrew Young (1807–1889). Composer of the hymn *There Is a Happy Land*.

Alan Brebner (d. 1890). Civil engineer.

Andrew Leslie (1818–1894). Shipbuilder.

Rev William Garden Blaikie (1820–1899). Writer and temperance reformer.

E.G.Elder (d.1918). Women's Royal Naval Service during Word War One.

E.W.L.Fruish (no date). Women's Royal Navy Service during World War Two.

Christian Salvessen (1827–1911). Norwegian merchant.

Thomas Peddie (1844–1911). Railway and civil engineer.

Thomas Aitken (1833–1912). Provost of Leith.

James Campbell Noble RSA (1846–1913). Artist.

M.P. Galloway (1843–1919). Shipbuilder.

George Ogilvy Reid (1851–1928). Artist.

Sydney Mitchell (1856–1930). Architect.

Monuments

Stones commemorating several members of the Stevenson family drowned in the Eyemouth disaster of 14 October 1881, when 189 fishermen died in a storm.

Newington

John Cairns (1818–1892). Minister.

General Sir Gordon Holmes MacMillan (1897–1986).

Edward Sang (1805–1890). Mathematician.

William Campbell (d.1902). Organist. Sculpture by William Grant Stevenson.

Arthur Lloyd (1839–1904). Music hall entertainer (grave lost).

William McTaggart (1835–1910). Artist.

Rev John Ross (1842–1915). Missionary.

Arnaud Massy (1877–1950). Golfer (grave lost).

Sir Charles Gibson Connell (1899–1985). President of the Scottish Wildlife Trust.

The following are a list of graveyards and cemeteries I have missed out, either because they are too modern to be included in a history of Edinburgh, but mainly because I had limited space.

Buccleuch Churchyard. Comely Bank Cemetery. Corstorphine Churchyard. Corstorphine Hill Cemetery. Craigmillar Castle Park Cemetery.

Cramond Churchyard. Currie Cemetery and Churchyard. Eastern Cemetery. Dalmeny Churchyard. Dalry Cemetery. Duddingston Churchyard. East Preston Street Cemetery. Gogar Churchyard. Kirkliston Cemetery. Liberton Cemetery and Kirk. Morningside Cemetery. Mount Vernon Cemetery. Piershill Cemetery. Sciennes Jewish Cemetery. Seafield Cemetery. There is also a dog cemetery in Edinburgh Castle.

References

Maldonado, A. *Rise of the Living dead: The Long Cist in Late Iron Age Scotland*. (2104).

M'Crie, Thomas. *The Life of John Knox* (Edinburgh, 1811).

Laing, David ed. *The Works of John Knox* (Edinburgh: Woodrow Society 1846–64).

Dawson, Jane E. A. *Oxford Dictionary of National Biography* (Oxford University Press 2004).

Lang, Andrew (1909). *Sir George Mackenzie, King's Advocate of Rosehaugh: His Life and Times* (London: Longmans, Green and Co 1909).

Lee, Sidney, ed. *Mackenzie, George. Dictionary of National Biography* (London: Smith, Elder & Co. 1893).

Scottish Covenanter Memorials Association.

Littlejohn, Henry. *Report on the Sanitary Condition of the City of Edinburgh* (Edinburgh 1863).

Keir, David. *The City of Edinburgh. The Third Statistical Account of Scotland* (1966).

Edinburgh Graveyards Project: *Documentary Survey For Canongate Kirkyard*.

Willsher, Betty. *Understanding Scottish Graveyards* (W&R Chambers Edinburgh 1985).

Willsher, Betty. *Midlothian and Edinburgh in Survey of Scottish Gravestones* (National Monuments Record for Scotland c.1985).

Rodgers, Reverend C. *Monuments and Monumental Inscriptions in Scotland* (1871). City of Edinburgh District Council. *Catalogue of Monuments and Burial Grounds* (1979).

Edinburgh Graveyards Project: *Documentary Survey For Canongate Kirkyard*.

Grant, James. *Cassell's Old and New Edinburgh : Its History, Its People, and Its Places* (London ; New York: Cassell, Petter, Galpin & Co., 1881).

Selby Wright, Robin. *An Illustrated Guide to the Canongate Kirk, Parish and Churchyard* (1965).

Turnbull, Mike. *The Edinburgh Graveyard Guide* (St Andrews Press 1991).

Dennison, Patricia. *Holyrood and Canongate: A Thousand Years of History* (2005).

Wright, Ronald Selby. *An Illustrated Guide to the Canongate Kirk, Parish and Churchyard* (1965).

Glendinning, Miles. *The Architecture of Scottish Government : From Kingship to Parliamentary Democracy* (Dundee: Dundee University Press, 2004).

Edinburgh Graveyards Project: *Documentary Survey For Greyfriars Kirkyard*.

Bryce, William Moir. *Old Greyfriars Church, Edinburgh* (1912).

Brown, James. *The Epitaphs and Monumental Inscriptions in Greyfriars Churchyard, Edinburgh.* (J

Moodie Miller, Edinburgh, 1867).

Smith, J. Stewart. *Historic Stones and Stories of Bygone Edinburgh* (1924).

Edinburgh Graveyards Project: *Documentary Survey For Old Calton Burial Ground*

Colvin, Howard. *A Biographical Dictionary of British Architects 1600–1840* (Yale University Press, 1995).

Anderson, William Pitcairn. *Silences That Speak. Records of Edinburgh's Ancient Churches and Burial Grounds, with Biographical Sketches of the Notables Who Rest There* (A. Brunton 1931).

Edinburgh Graveyards Project: *Documentary Survey For New Calton Burial Ground.*

Gray, W Forbes. *Historic Churches of Edinburgh* (Edinburgh & London: The Moray Press, 1940).

Edinburgh Graveyards Project: *Documentary Survey For St Cuthbert's Kirkyard.*

Birrell, J.F. *An Edinburgh Alphabet* (Mercat Press, Edinburgh 1980).

Royle, Trevor. *Precipitous City* (Mainstream Publishers, Edinburgh 1980).

Daiches, David. *Edinburgh* (Granada Publishing 1978).

Cole, Hubert. *Things for the Surgeon* (Heinemann, London 1964).

Lochhead, Marion. *Edinburgh Lore and Legend* (St Edmundsbury Press 1986).

Morris, Albert. Kay's Capital Characters (Pentland Associates 1996).

Brice, William Moir. *History of the Old Greyfriars Church Edinburgh* (William Green and Sons 1912).

Smellie, Alexander. *Men of the Covenant* (Andrew Melrose, London 1908).

Willsher, Betty and Hunter, Doreen. *Stones: 18th Century Scottish Gravestones* (Canongate Edinburgh 1978).

Weever, J. *Discourse of Funerall Monuments* (London 1631).

Fawcett, Richard. *The Palace of Holyroodhouse: official guide* (HMSO, Edinburgh, 1988).

Acknowledgements

Thanks to Charlotte Corstorphine, Ali Bowen, David Connelly, Eve Harvey, Charlie Guy and the late Jim Gilhooley for the extra information.

Jan–Andrew Henderson is the author of the non-fiction books *The Town Below the Ground: Edinburgh's Legendary Underground City*, *The Emperor's New Kilt: The Two Secret Histories of Scotland* and *The Ghost That Haunted Itself: The Story of the Mackenzie Poltergeist*. He is also a children's novelist, whose last three teenage thrillers were short listed for thirteen literary awards and won the Royal Mail and Doncaster Book Prizes. janandrewhenderson.com